A Guide to Academia

Getting into and Surviving
Grad School, Postdocs,
and a Research Job

A Guide to Academia

Getting into and Surviving
Grad School, Postdocs,
and a Research Job

Prosanta Chakrabarty, Ph.D.

A John Wiley & Sons, Ltd., Publication

Wiley-Blackwell is an imprint of John Wiley & Sons, formed by the merger of Wiley's global Scientific, Technical and Medical business with Blackwell Publishing.

Registered office: John Wiley & Sons, Ltd, The Atrium, Southern Gate, Chichester, West Sussex, PO19 8SQ, UK

Editorial offices: 2121 State Avenue, Ames, Iowa 50014-8300, USA
The Atrium, Southern Gate, Chichester, West Sussex, PO19 8SQ, UK
9600 Garsington Road, Oxford, OX4 2DQ, UK

For details of our global editorial offices, for customer services and for information about how to apply for permission to reuse the copyright material in this book please see our website at www.wiley.com/wiley-blackwell.

Library of Congress Cataloging-in-Publication Data

Chakrabarty, Prosanta.
   A guide to academia : getting into and surviving grad school, postdocs, and a research job / Prosanta Chakrabarty, Ph.D.
      pages cm
   Includes bibliographical references and index.
   ISBN 978-0-470-96041-7 (pbk. : alk. paper)
   1. Universities and colleges–Graduate work–Handbooks, manuals, etc.   I. Title.
   LB2371.C47 2012
   378.1'55–dc23

                                                         2011048156

A catalogue record for this book is available from the British Library.

Wiley also publishes its books in a variety of electronic formats. Some content that appears in print may not be available in electronic books.

Set in 9.5/12.5pt Palatino by Aptara® Inc., New Delhi, India
Printed and bound in Singapore by Markono Print Media Pte Ltd

1   2012

Dedication

To those students who want to continue learning and giving back. . .

Contents

Preface

I've tried to write the most straightforward guide to the academic life as possible, without the frills of first-hand accounts. (Except in those few cases where these accounts can help clarify the advice.) This book is certainly not a memoir, but it is my singular vision as it was shaped by my experience and the advice I've received over the years. As I've gone through the academic process, I've made mistakes, and this book is my attempt to help budding academics make the right decisions in order to efficiently reach their goals without making those same mistakes.

I started thinking about writing this book around my third year in graduate school. This was about the halfway point of my Ph.D., and I was right in the middle of older graduate students who were too busy writing up their work, and "newbies" still getting their footing. I was in a position to give advice for the first time, having advanced to candidacy and having already seen others pass and fail. Initially, this text was going to be a guide for graduate students, but alas, the science came calling and there was no time for extras. I am now an assistant professor, and even though there is even less time for extras, I've picked up on this idea again, expanding it over several professional levels with a heavy focus still on the graduate stage. This book can help the inquisitive undergraduate who is considering graduate work, but this text will also help the newly hired assistant professor struggling with the pressures of tenure. By seeing the entire academic landscape laid out before you, the reader can gain a sense of perspective no matter the current point of your career.

I know I've had some luck along the way (some would say a lot of luck), but I've learned some things that I think will help others traverse the academic wilderness. As with all advice, the reader will need to tailor it to fit his or her own particular circumstances.

Each stage of your academic career is about getting to the next stage. You should realize that you are always being judged. The judges are your academic advisors, colleagues, and peers. This book will help you through the process of knowing what is expected of you and how to excel during the process. I've written this guidebook to focus on those who wish to have research academic careers. Engineering students and medical students have their own set of particular loopholes, and they perhaps will not get as much out of this guidebook as others.

Each person will have a different graduate experience; this guide is one that will help you through the general process and to learn what to expect. You should also talk to other students and professors and see how they got to where they are to fine-tune your guidance. Everyone ultimately has to pay their dues to get to where they want to be. There is no easy road to your goal; if there is, your goal is not ambitious enough. As you struggle through those rough patches, keep in mind what still needs to be accomplished to reach your destination. It may seem daunting, but hopefully with the help of this book, the steps you need to reach your academic goal won't come as a surprise.

During every step of the way, you should make time to talk to people and get their advice about moving along in the process. One of my favorite words of encouragement about the academic process was, "Everyone eventually ends up in the right place." Remember those little words of wisdom and encouragement. I did, and that is largely what makes up this book.

This book is what I wish I had been handed when I first started thinking of becoming a scientist, and I hope it can save the reader some heartache and growing pains as you start on your own path to your ideal career. Good luck to you, and happy reading.

Prosanta Chakrabarty
Baton Rouge, Louisiana
November 2011

Acknowledgments

To my many academic mentors—Melanie Stiassny, Bill Fink, John Sparks, Ellen Pikitch, Jerry Smith, Chris Austin, Robb Brumfield, James Albert, Bryan Carstens, Ron Oldfield, Deborah Goldberg, Doug Futuyma, Ian Harrison, John McCormack, Marc Ammerlann, Scott Schaefer, and countless others—thank you for showing me through your experience and shared wisdom what it is to be a scientist and academic. I hope you see your many lessons retold in these pages.

To my parents, Chitta and Anurupa Chakrabarty, who always told me to do what I love—thank you for being the first to teach me how to teach and to learn. To my wife, Annemarie Noël, whose support and understanding allow me to accomplish everything in my life from getting dressed in the morning to writing this book. As they say, "behind every great man there is a great women," but of course there is a surplus of great women so sometimes a regular dope like me gets lucky and snags a great woman, too. My twin daughters Chaya and Anjali were born between the first and second drafts of this book, and their arrival gave me the courage to write the last chapter, which I dedicate to them. I hope by the time they choose their career path, they won't have to worry about gender inequality.

Many thanks to the Wiley-Blackwell team. I especially would like to thank Justin Jeffryes who sought me out, signed me up, encouraged me, and fought on my behalf. I would like to thank Shikha Sharma for her tireless efforts in the copyediting and proofreading of this book. I must also thank those who reviewed different sections of the manuscript and provided valuable comments, including numerous anonymous reviewers but also Parker House, Elizabeth Derryberry, Jahi Chappell, and Jeremy Wright. I must also thank my lab—Caleb McMahan, Wilfredo Matamoros, Matthew Davis, and Valerie Derouen—who (without their knowledge or consent) I have been secretly reading this book to under the guise of giving advice.

A Guide to Academia

Getting into and Surviving
Grad School, Postdocs,
and a Research Job

1 Preparing for Entering Academia

The Hard Truth About the Academic Life

Choosing the academic life is not an easy choice to make. But, if you feel it is the right choice for you and you are a hard, smart worker, then it can lead to a permanent position in a job that is intellectually fulfilling, stable, and rewarding. The hurdles to obtaining an academic research job (a professorship or equivalent) include getting the following: a bachelor's degree, research experience, possibly a master's degree, a doctoral degree (Ph.D.), maybe 2–5 years of postdoctoral research experience, an academic job (e.g., assistant professorship), and tenure (Figure 1.1).

You might be able to skip a few of these steps; perhaps you won't need a master's degree or a second postdoc, but there are still many hurdles that you will need to overcome to achieve your goal. Luckily these "hurdles" are what trains you to be an independent researcher and thinker, and that training can be among the most fulfilling and worthwhile experiences in your life. There is no real "typical path"; every individual's experience will be different. Academia is fun; you basically get to do what interests you, discover new things, and interact with other people who are also having fun learning about things they are interested in. It isn't always a bed of roses, of course, but compared with some other ways of making a living, it is a great gig. One academic I know always replies to the question "How's your job?" with "Beats workin'."

A Guide to Academia: Getting into and Surviving Grad School, Postdocs, and a Research Job,
First Edition. Prosanta Chakrabarty.
© 2012 John Wiley & Sons, Inc. Published 2012 by John Wiley & Sons, Inc.

LONG PATH Tenure

4 yrs	2–3 yrs	4–7 yrs	1.5–3 yrs	1.5–3 yrs	5 yrs	5 yrs	5 yrs
Bachelor's Degree	Master's Degree	Ph.D.	Postdoc 1	Postdoc 2	Assistant Professor	Associate Professor	Full Professor

AVERAGE PATH

Bachelor's Degree	Master's Degree	Ph.D.	Postdoc 1	Assistant Professor	Associate Professor	Full Professor

SHORT PATH

Bachelor's Degree	Ph.D.	Assistant Professor	Associate Professor	Full Professor

Figure 1.1 **The various route's to academic enlightenment. Dark bars are positions that are necessary and light bars are positions that are not always required.**

The academic life has a lot of perks and can be very fulfilling, but it might not make you rich. Starting annual salaries range from $40K to $80K, and most never make it much past the $100K mark. However, there is a great deal of job stability, and—best of all—if you succeed, you can make a living doing something you enjoy. Academia is extremely fulfilling to those who find satisfaction in solving problems in a particular area of research. Researching subjects that interest you, teaching what you know best, and, for the most part, making your own schedule are among the freedoms that research academics have that few others can claim. You can work 9 to 5 if you like, or 5 to 9. Nobody will be looking over your shoulder, telling you what to do; you must be your own nagging boss. Your ambitions have to match your academic goals; without this motivation, you will undoubtedly fail. You will receive very few pats on the back in academia; the only way you know you are doing a good job is if you notice that fewer people are complaining. Be warned, academia is not for the faint of heart. You will submit papers and grant proposals that you think are the best things since sliced bread, and you will be shocked by reviewers and advisors who will knock you down as low as you can possibly feel. The rewards of success are great, but only because academic success (e.g., discovering a new species, falsifying a long-standing hypothesis, etc.) goes hand in hand with your happiness. If you don't get excited about going to work for science, then science will make it very hard for you to succeed.

The people who I see achieving the most success are not necessarily the smartest, but they are almost always the hardest working. You might want to go to graduate school in order to become a lecturer at a small liberal arts college with no research component, but you will still have to play the game with those students getting doctoral degrees in order to land research positions at top universities; you won't be treated differently just because you have different goals and ambitions.

Academia is all about self-motivation, but one great incentive is tenure. (Tenured professors are those who have their positions guaranteed [i.e., they can't be fired unless they do something prohibited or illegal]; assistant professors are essentially on probation.) Tenure gives you nearly complete academic freedom. Getting tenure means you have made it through the academic ringer and that your institution wants to keep you forever. You will only be granted tenure once the institution is sure that you are self-motivated enough to keep working hard, if not for them, for your own sake. If for no other reason, your work should keep your interest; as the old adage goes, "If you love your job, you never have to work a day in your life."

Getting a Head Start as an Undergraduate Volunteer

It is never too early to start building up your resume by getting research experience. Research experience is something you most certainly need to distinguish yourself among the hoards of undergraduates whose only experience is classwork. This experience will also tell you how much you like research, and the more research you do, the more fun and interesting the research you are offered becomes. If you volunteer to work for an academic, you might start by doing menial jobs, like washing lab equipment or filling boxes with neatly arranged pipette tips. Once you've proven to be someone who can be trusted to arrive on time and not break things, you may then be asked to do something more exciting, like mixing and preparing chemicals or taking x-rays. This kind of menial experience is actually a good start. Begin creating your CV and adding things like the following:

> *Volunteer*—Wainwright lab, University of California Davis. Made x-rays of fishes for studies on the ecological morphology of darters (Etheostomatinae). Fall 2011

This might sound obscure, but someone looking at your CV might know Peter Wainwright, or they may need someone who knows how to make x-rays, or who is interested in darters. You want to provide as much information as you can in just a few lines (see more about your CV below).

You might not be the only volunteer in the lab; you might see that other volunteers are doing things you wish you were doing instead: Be patient, your time will come. People who work hard, learn quickly, and are inquisitive are almost always offered more opportunities. Whiners, complainers, latecomers, and the generally unenthused are quickly shuffled out the door.

You might not work directly with the PI (principal investigator, or head of the lab); more often you will work for a graduate student or postdoctoral fellow. You might even end up getting paid for your work, which is great. More important than the money, though, if you can believe it, is the experience. The ultimate prize is to be included as an author of a publication. A research publication with your name on it means you have just joined the ranks of researchers. People who Google you (or better yet, Google Scholar you) will find your paper and be impressed, and you will have an entry in the most important category of your academic CV—*Publications*. Before you get to that point, however, and maybe before you even get to work on a real project, you might have to wash 1,000 dirty beakers in a sink for 8 hours a week for 3 months.

If you volunteer and are excited about working in a lab or on a research project, that's great. It is okay to be happy, gregarious, and inquisitive. There is a fine line, however, between being outgoing and being annoying. If you are working with someone who is writing a manuscript on the computer in the other room, don't just barge in and start chatting them up about the party you went to last night. Academics are busy people with a lot on their plates, and they typically have short periods of time in which they need to concentrate fully on a particular task. Breaking their concentration with a myriad of questions will not go over well, but having all of your questions answered in one shot is better than interrupting someone five times in 1 hour. Try to gauge your advisor's reactions to your questions to see whether he is becoming frustrated. It might not be your fault if he is upset, but it can quickly become your fault. The best volunteers are problem solvers and note takers who understand their roles and who want to move up in the world and know how to get there.

Try to solve problems yourself but not at the expense of making a mistake that could end up costing a researcher even more of his precious time. If the PI keeps his door closed, it is probably closed for a reason. But, an emergency is an emergency, and if you have a question that can stave off an emergency you better ask it—closed door or not. However, if there is a graduate student or technician you can ask instead, you probably should. As with any job, fitting into the group's social dynamics is almost as important as the work itself. Remember that, as a volunteer, you are on the lowest rung of the totem pole. That doesn't mean you should be mistreated, but it does mean you may not have as much access to the PI's time as you would like.

Aside from being including in a publication, the other thing that can be more valuable to you than money is a nice recommendation letter. Your professors or lecturers might write you recommendation letters for graduate school because you got an "A" in their classes, but these are typically informal letters that they write dozens of every semester. They might do little more than replace the name of the last "A" student who asked for a recommendation letter with yours. (You'd be surprised how many letters I've read where the gender of the student is incorrect throughout the recommendation, because the author didn't bother changing the gender along with the name.) If you work for someone as a volunteer for a couple of months, you should most certainly ask for a recommendation. Even if you worked most closely with a graduate student or a postdoc, you should ask them to write a letter that is signed by the PI as well. The person who wrote the recommendation letter, in most cases, carries more weight than the actual content of the letter. If you get a recommendation letter from E.O. Wilson or Stephen Hawking, it won't matter if it is two sentences long; it is still better than a 15-page letter written by your graduate student TA (teaching assistant) who only knows you from class.

The Undergraduate CV

The CV is your *curriculum vitae*, from the Latin for "course of life." It is your introduction to all who you will meet for whom you want to work and impress. If you send someone an e-mail about wanting to work for them, the first thing they will do is look you up on the Internet. If your Facebook profile is a picture of you drunk and naked at a frat party, then your correspondence with that person will likely not go any further than your introduction. Aside from being diligent about your social media profiles, you should also write a professional CV that you would use in corresponding with academics.

As an undergraduate, you are not likely to have a 15-page CV, but that's okay. The important things you need to showcase are your skills, who you've worked for, and your other relevant experience. See Appendix 1 for an example of an undergraduate CV.

Finding Your Research Interests

Perhaps you might only have a vague idea of what you are interested in—that's okay. If you know you like quantum mechanics and you admire Schrödinger, at least that's a start. It is okay to have only a broad interest at first. Don't think about your interests in terms of what jobs are out there or how much money you'll make; instead, think only of spending your life studying something that you are interested in. Over the course of your undergraduate career, you met interesting professors and learned about interesting things from your textbooks and friends. Certainly some area of research piqued your interest; otherwise, you wouldn't be reading this book. Perhaps you were listening to a lecture in a natural history class and thought, "You mean they still haven't figured out how orcas communicate in a hunt?" If you want to find out more, there is of course the library and the Internet. But- to find out what really is going on you need to start reading the primary literature, that is, science journal articles. Learn how to use Google Scholar (http://www.scholar.google.com), where you can type in a subject or a researcher's name to find available articles. Also learn how to use the ISI Web of Knowledge, or Scopus, which are the grown-up versions of Google Scholar. If you don't know how to use these or don't have access to them, ask a professor, librarian, or graduate student who does. Once you've found articles that interest you, look up more papers from the citations in the Literature Cited or look up more papers by the same authors. If you find yourself looking up work by the same authors, you might want to contact those authors as potential future advisors. Just as with food, only you know what you like. Discovering what you like most will mean sampling lots of things. If you want a career in science, you will have to read a lot. The more you read, the more you'll understand and the better prepared you'll be for the next step. The earlier you know what you like the most, the quicker you can get started on focusing on learning all there is to know on that subject.

Summary

- Although the academic life may not be the most glamorous, and it won't make you rich, it can be incredibly fulfilling.
- Self-motivation is the key to academic success. Because few people will give you encouragement, you will have to really want to accomplish your academic goals because you are genuinely interested in discovery and learning.
- Being a tenured academic is a sweet deal, and few other jobs offer the same kind of flexibility and stability.
- Moving up the academic ladder will mean that you have to leave the ground. Start by volunteering in labs that you admire, and begin building a set of skills and an array of experiences that will help you gain more research experience
- As you gain experience, start creating a CV that you can use as your "introduction" when contacting academic professionals.
- Get recommendation letters from established PIs rather than from PI's graduate students or postdocs.
- Read scientific articles that interest you to narrow your focus on the research that most interests you.

2 Applying for Graduate School

The Hard Truth About Applying to Grad Schools

You know you need a Ph.D. to get to where you want to be in life; do you have the grades and the background to get into a Ph.D. program? Entering a graduate program isn't like getting into college for a bachelor's degree. The admissions committee will look at your grades and standardized test scores, but these play a relatively minimal role in the selection of a graduate student.

Four target areas to help you self-evaluate your chances of getting into a graduate school include the following: (1) GPA—Typically you want to aim for at least a 3.4 undergraduate GPA, with an even higher GPA in classes that are in your major, at least if your major is the same subject for which you want to go to graduate school. Hopefully that is the case, but if your major was chemistry and you want to go to graduate school in biology, you might have to show that you have enough of a background to be a good biology graduate student. (2) Classwork—Although you can

---

A Guide to Academia: Getting into and Surviving Grad School, Postdocs, and a Research Job,
First Edition. Prosanta Chakrabarty.
© 2012 John Wiley & Sons, Inc. Published 2012 by John Wiley & Sons, Inc.

take classes as a graduate student, you generally want to avoid having a heavy course load so that you can focus on your graduate research. For this reason, it is good to show that you have some graduate level courses on your CV and courses that generally increase your background knowledge in the subject you want to study. For instance, if you want to be an Evolutionary Biologist, you should definitely take undergrad classes like Evolution, but you should also see if you could get into seminar classes and reading groups in upper division/graduate courses that are offered. Ask your professors if they teach any seminars and if they think you could be added to the course. Joining reading groups and journal clubs with graduate students is great experience, but it will also look good on your application for graduate school. (3) GRE scores—The minimum score on your GRE (Graduate Record Examination) required for acceptance into graduate school varies widely depending on the institution you are applying to. If you get a score below the minimum "requirement," that doesn't necessarily mean you are out of the running. It also doesn't mean that these scores don't matter. Your GRE scores might not be evaluated in the way that you think. Because the GRE is a standardized test, it is the only category in which all candidates can be judged on a level playing field. If all else is equal between two candidates, then GRE scores might be used as a tiebreaker. In addition, GRE scores often determine who among the applicants will receive graduate scholarships (see more about GREs below). (4) Research experience—If you have no research experience you will have a hard time getting into graduate school. If you want to get into a Ph.D. program, you almost certainly need research experience and probably even a publication or two. The more research experience you have, the better. Without any research experience it will be difficult for graduate programs to evaluate how you will do as an independent researcher. Even if you have great GRE scores and a very high GPA, if you have no research experience, then your ability to perform research is unproven.

If you are lacking in one or more of the above four criteria, don't fret too much; any of the above can be fixed. What is most important is your desire to become an academic. If you want it badly enough and you are willing to work to achieve your goal, then you will most certainly achieve that goal. It may seem like that goal is so far away that it isn't worth trying or that it will take too long to get there, but you have to put things into perspective: Life is long; a few years of training and sacrifice won't seem so bad once you've gone through it. You don't want to be asking yourself whether you gave your dream your best shot once the opportunity has passed. Instead of spending the next 5–10 years or so getting a Ph.D. and doing a postdoc, you could take some time off and see what else is out there; but those years will still pass by, whether you are achieving your goals or not. You will have to ask yourself what you ultimately want to do and what sacrifices you are willing to make to get there.

Master's Degree versus Ph.D.

If you know you want to be a scientist or a career academic and you have the research background to skip a master's degree, then I would recommend doing just that. Some students fool themselves into thinking that they are not ready for a Ph.D., even though they have the right credentials. A Ph.D. thesis is essentially a lengthy master's (4–6 years, instead of 2–3 years). Some master's programs are more like extensions of a bachelor's degree, in that the student is just taking classes. Students with this kind of master's degree, one with little independent research, will have a hard time entering a research-oriented Ph.D. program. What Ph.D. admissions committees are looking for is a proven ability to do independent research and a match with a willing advisor.

Some students may have no choice but to pursue a master's degree. If you are unable to get into a Ph.D. program, then a master's degree is a great fallback and a way to gain some research experience. Many Ph.D. advisors will not take students unless they have a master's degree already. In some cases, a master's student can transition that degree into a Ph.D. at the same institution, rolling the efforts toward the lesser degree into the beginning efforts of a doctorate. This can generally be done by petitioning your advisory graduate committee into allowing for a more ambitious and lengthy project that would fulfill the requirements of a Ph.D. However, even with this switch, you will oftentimes still need to formally apply to the Ph.D. program.

All top-tier universities (i.e., research I universities) have a Ph.D. program, as do many mid-range to small universities. If there is a Ph.D. program at a university, then the master's students are typically treated as second-class citizens relative to those receiving a doctorate. Look at it from the advisor's perspective: The time to train a master's student is not that different from that of a Ph.D. student, perhaps 1–2 years of intense training; the difference is that the master's student will typically join a new lab elsewhere with the experience they just received. So from the advisor's view, it is worth investing in the Ph.D. student who will remain in the same lab in which they were trained. Master's students often have to fight to find teaching assistantships and other funding sources that are usually guaranteed in Ph.D. programs. At many schools, master's students still have to pay off their tuition just as they did with their bachelor's degree, whereas Ph.D. students are typically paid a stipend and have a tuition waiver. Master's students will often also be expected to take on the role of TA and take classes, all while trying to complete their own research. Many master's programs even require their students to take as much as twice as many classes as students in the Ph.D. program, even though the lesser degree usually takes half of the time. (This is often to make sure the master's program pays for itself: credits = $.)

There are many negatives to a master's program relative to a Ph.D. program, but if you have no choice, then there are many benefits to pursuing a master's degree. A master's degree will certainly give you a leg up as a candidate for a Ph.D. program. A student applying for a Ph.D. with a master's degree, some publications, and independent research experience is much more sought after than an unproven undergraduate. However, this means you should complete a master's degree with a thesis option. If you are only taking classes, then your experience is just an extended bachelor's degree.

I have often found that master's students at universities without a Ph.D. program are better students than master's students at Ph.D.–granting institutions. At these schools that lack a Ph.D. program, undergraduates who want to continue on to graduate school, but lack enough research experience to move on, stay on at their undergraduate institutions and shift from a bachelor's to a master's program. There, they can be nurtured by familiar teachers in projects they often started as undergraduates. Many of these students flourish and graduate with a well-rounded project. At a Ph.D.–granting institution, master's students often flounder under the competition with Ph.D. candidates, who are given preferential treatment for the limited time and attention of advisors and funds.

Apply to People not Programs; and Finding a Good Fit

The approach to applying for graduate schools is not the same as applying for your undergraduate education. Everyone wants to be associated with great universities, such as Harvard, Oxford, or MIT, but the research path you choose may not exist at these institutions. You should fine-tune your search for graduate schools by searching for potential advisors. If you want to study plant physiology, you will be wasting your money by applying to a university that only has vertebrate biologists. You might see that Harvard has a great evolutionary biology program, but unless there is a professor there who matches your exact interests, then there is little chance you will get in. Someone at the institution you are applying to needs to select you as a student; the best chance of this happening is if they see that your research interests match theirs. You want to do a search of researchers in your general field of interest: developmental biology, systematics, etc. By the time you start applying for schools, you should have read papers by scientists you admire. E-mail these scientists and ask them if they have space in their lab for someone like you. The right and wrong ways of composing these introductory e-mails will be discussed later in this chapter.

A good fit with an advisor means a lot more than a high undergraduate GPA and high GRE scores combined. The number of people admitted into a graduate program is a small percentage of the applicants; unless someone is fighting for you, you'll have a hard time getting admitted.

So how do you find a good fit? If you have an interest in a particular subject, read some recently published articles and books on that subject. What portions do you find most interesting? Which ones can you imagine working on? Try to think of a project on that subject, and imagine several stages of analysis from narrow to broad. Do your best to think of a proposal for a Ph.D. or master's project on that subject. Study most closely articles written by people who you might be interested in working with. Check their websites and institutions, and when you feel comfortable about the subject, contact them. A good start to contacting people is saying that you've read their articles and that you are interested in working along a similar path.

Sometimes when looking for an advisor, you may find that the established folks are nearing retirement and not taking students. In that case, I would look at the former students of that professor; those students themselves may now be established faculty and perhaps more attentive advisors. The diversity of students is matched by the diversity of faculty. Some students need a lot of handholding and a lot of faculty are very hands-on; some students are very independent (these are generally seen as better students to have) and some faculty are very hands-off. A good fit is what you want; you should contact former and current students to solicit their opinions about a potential advisor. A bad fit in graduate school can seriously derail your career.

You should always take an opportunity to meet your potential advisor. A face-to-face meeting or, at the very least, a phone call or Skype conversation will help you feel each other out. The one thing you don't want to hear during these conversations is your potential advisor complaining about how annoying his students are and how little the students have done for him. The one thing they don't want to hear is a student complaining about his current situation and how hard it is to get along with certain people. The advisor wants to hear from a potential student with research interests similar to his own, and he wants to see how the student can apply the skills they have to tackle those common interests in a new way. Research faculty want students to infuse new ideas and excitement to a lab, adding a new perspective to old questions, or to add new questions and directions to the lab.

During an interview, ask what the advisor wants out of a student. You don't want to be a Ph.D. student who is little more than a lab technician for a PI, essentially following the command of that PI to carry out his research goals without your input. That kind of student will not learn the independence needed to become an academic. You want to be trained to one day be in a position to conduct your own work and come up with your own program of study. If you are just going to be running experiments anonymously in someone's lab, then you will not learn what it takes to be part of the scientific community. You want to join a lab where you will be nurtured and taught, where you can write your own papers and give presentations that are from your own ideas.

One other important thing to keep in mind when looking for the right graduate program is whether you will be earning enough to get by and whether you will have enough research funding. Graduate students get paid very little ($18K to $25K a year is typical); it is enough to survive, but you should make sure that you will be able to live somewhat comfortably, particularly if you have a family. Make sure your health insurance will be covered as well (it generally is). Sometimes you do have to make sacrifices to work with someone you really think will take you where you want to be, but this should not come at the cost of your health. You should also ask your potential advisor what kinds of funds are available to support graduate student research. There is little you can do as a graduate student without the funds to conduct research, so inquire about the kinds of internal (within university) funding sources available, or whether the advisor has funding. No matter how brilliant someone is as a scientist, if there is no money to conduct research, the research won't get done. The graduate advisor should also know of external sources from experience with previous graduate students; if they don't, they might be out of touch. If you know that you need funds for certain equipment to carry out your research, don't be afraid to bring this up during an interview. It won't just make you sound smart, but you will learn quickly how in touch your advisor will be with your needs and plans.

You should also ask about the required course load for graduate students. For some programs, students are required to take as many as two full years of classes; at others, no real classwork is required. The difference between these two can be the difference between a Ph.D. that can be earned in less than five years and one that takes more than seven years. Ask what the average time is to complete the degree for people in the program. If you think you can jump directly into research, go for a program that has fewer required courses. If you know you would like to conduct fieldwork for six months out of the year and can't take classes for a while, ask if that would be an issue. At some schools, many of the required credit hours can be taken as independent research hours. If you are a biology student and your program asks all Ph.D. candidates to take microeconomics and chemical engineering classes, you may want to rethink applying to that program. You don't want to be forced into courses that aren't training you in your field of study.

Contacting Professors

When inquiring about entering someone's lab, you should establish contact with your potential advisor to gauge his or her interest. When sending an e-mail, remember that the professor is a very busy person who gets dozens of e-mails a day from people asking for something. These advisors have to filter through dozens of e-mails from potential students every week. If you contact an academic you do not know, write as formal an e-mail

---

Box 2.1   An example of a bad introductory letter

Hello, I am interested in the PhD program in ichthyology. I received my bachelor's degree from XXXXX University. I graduated in 2008. I decided to take a short time off before acquiring my PhD. If you could please send me some information regarding the program at LSU I would appreciate it.

Thanks,
Space cadet

---

as possible. An example of a good letter has all the information that the potential advisor will need if he wishes to contact you again. This required information includes a CV and a summary statement of research goals.

An example of a bad letter I received recently is shown in Box 2.1. This letter tells me absolutely nothing about the person, except they went to university and they think it is easy to get into a Ph.D. program. It also tells me that they think I will have so much time on my hands to answer their ambiguous e-mail.

Compare this with a good letter (Box 2.2):

Box 2.2   An example of a good introductory letter

Dear Sir,

I am XXXXXX, a graduate student of XXXXX, presently completing a M.Sc. degree in Ichthyology at XXXXX University. I have read some of your papers and your website and I am greatly interested in continuing my education with you. I would like to carry out systematic studies on fishes. I have interests in working on a number of marine and freshwater groups.

Zoology has always been my best subject and I find greater pleasure working with animals and nature. I have recently completed a research project titled "The biogeographic history of fish assemblages on Madagascar". This draft manuscript is attached.

I have published two papers: "A taxonomic revision of XXXX" in Zootaxa (2008) and "A description of a new species of XXXX" in Fish Biology (2009). I have also been awarded the XXXX Memorial Scholarship. These two papers are also attached.

I have summarized my interests and qualifications to pursue a Ph.D. in my CV that you can find on-line at www.mywebsite.net.com. With my training and background as a zoology student, my experience as a researcher and a teacher, I am confident that I will be an excellent addition to your lab.

Yours sincerely,
Serious-about-school

The difference between these letters is night and day. I get a few of these a month; guess which type I focus on and which I ignore?

The Graduate Application

Now that you've decided on a course of study, and to whom you want to apply, you need to formally apply to the university's graduate program. Typically, I wouldn't recommend wasting your money applying to a particular university until a professor you contacted showed interest in accepting you into his lab. The first thing an admissions committee will look for is evidence of a connection between a candidate and a potential advisor. If a faculty member wants a particular candidate and that candidate wants to work with that faculty member and has the right credentials, then that candidate is typically given an offer.

The admissions committee at a top research institute may look at more than 100 applications at one time, so yours needs to stand out from the crowd. Even a great recommendation from a potential advisor isn't enough to insure that you will get accepted. You will also need good research experience, good recommendation letters, and relatively high GPA and GRE scores. To put you over the top, published papers, patents, presentations, and other evidence that you will be a good student are vital to getting into the program you want.

When the admissions committee is reviewing the applications, they look at the completed applications first. If there are some application packages missing recommendation letters or GRE scores, they will not be judged at the same time as the majority of applicants. Some committee members may feel that, if the candidate was unable to meet this deadline, then that is a good sign that the candidate would make a poor graduate student.

What usually keeps a package from being complete are recommendation letters. These are perhaps the most annoying things for a candidate to get. You will have to ask professors, some that you don't know that well, for letters, sometimes many letters (one for each application), on your behalf. Sometimes they forget or are tardy. The best way to avoid these situations is to ask for them to give you a signed and sealed letter that you can include in your application packet. This way you can remind them (i.e., badger) until you get the letter in hand. However, be aware that many programs now ask for these letters to be submitted on-line.

Don't ask a TA or a graduate student for a letter unless you are really desperate; a letter from a faculty member is always best. You don't get to read these letters, so make sure you get them from someone you trust to write a positive letter. Short letters are okay, but admissions committees are always scanning these for a negative word. Even so much as a "he sometimes has trouble working with others" or "she needs hands on mentoring" can kill an application. Typically, lecturers and graduate students with less experience writing these letters will feel the need to write something negative, just so the letter doesn't sound overly positive.

Statement Letter

Perhaps the most important thing in your application package is your "letter of intent," or statement letter (see Box 2.3). This is your chance to tell the committee and your potential advisor what you would like to do and what your goals are in your own words. A statement letter that reads like a Ph.D. proposal plays much better than one that is an airy personal statement about why you always wanted to be a scientist. I am always surprised by how many applications start off with some variation of "I've wanted to study XXXX since I was five years old." The second most

---

Box 2.3   An example of a good application letter

Dear Admissions Committee,

I am very excited about being a graduate student in your department. I have contacted Professor X and found that she is a great fit for my research interests. My interests in *blah, blah, blah* closely match hers. I have done some work on this subject in the lab of Dr. XXX (see recommendation letter and CV). This work has resulted in a publication in *Science*.

I am excited about continuing research in the field of *blah, blah, blah*. It is my hope that one day I can become an academic: teaching and doing research on *bla, bla, bla*. I recently read some articles by Drs. X and Y and thought that their research on *blah, blah, blah* was very interesting and that their results would take this field in a new direction by *[something clever showing your understanding]*.

I was delighted to find that your faculty includes Drs. G and K. I think these two excellent researchers would make great Ph.D. committee members for me because of their research interests in *yada, yada, yada*. I think they could greatly improve my research. I know my research experience is still limited, but I think I have the patience to learn the experimental techniques of *HHHH* and *SSSSS* relatively quickly and jump into independent research.

Please let me know if you require further information about my past projects or about me personally. My recommendation letters from Drs. A, B, and C are enclosed, as is my CV, a copy of my publication, and the remainder of my package. I hope that you see me as a strong candidate and can give me a chance to pursue my dream of studying *blah, blah, blah*. I hope I have the pleasure of meeting you during your recruitment weekend.

Sincerely,
The excellent recruit

common opening sentence is some variant of "When I was in the Amazon rainforest staring at a scarlet macaw last month all I could think about was how I would want to do this for a living." These aren't necessarily bad ways to start a letter, just hackneyed; however, if the rest of the letter is more personal statements and proclamations, instead of some assemblage of concrete goals and evidence of skills, it will be hard for the committee to gauge why you would make a good graduate student. Again, the best letters of intent read like a proposal for a research project. It is rare that a student can come in knowing exactly what he wants to do, but the more details you can provide, the better. A good approach for someone with limited experience is to mention methods you would like to learn in order to accomplish some scientific goals. Use journal article references in your statement to show that you know the literature and that you can write intelligently on a subject.

Play up any major research experience that you have. If you have a manuscript published (or in the works), mention the details of what you did and the interesting results you found. Despite all you may want to discuss in these letters, I recommend keeping them short, perhaps two to three pages. You can refer to your CV if you want to show the committee specifics about the range of work you have done.

Your personal statement is not a place for filler. No one needs to know about your family or how you once had a pet lizard, or if you survived a traumatic experience. These things only distract from your research experience and knowledge of a subject. Remember, the admissions committee wants to bring in people who have the best chances of succeeding in the Ph.D. program and who will go on to have successful careers.

You should name-drop some faculty that you think you would like to work with, mention that you contacted them and received some positive comments to make your statement letter stand out. If you mention faculty other than your main potential advisor, mention them as people that may be good members of your graduate evaluation committee because they share similar interests with you. This will show that you did your research. Name-dropping also helps the committee remind those people you've mentioned that there is an application they might be interested in reading, this will help build support for your application. Name-dropping too many people or people you don't have a sincere interest in working with can count against you. The faculty you mention should be people you genuinely would work with; if you are naming names just for the sake of it, you will come across as unfocused.

Send a copy of your final draft application letter via e-mail to potential advisors and ask them to send you comments before you send in a finalized copy. (Make sure to check grammar and spelling!) Your potential advisor might be too busy to look through it, but don't be discouraged. If you send the advisor this statement, you will give him some notes for writing a letter of recommendation. Often, the admissions committee will require the professor you are interested in working with to write a letter to put in the application package. Typically, this letter is a positive statement that the advisor thinks highly of the applicant.

Don't wait until the last minute to submit your application. As I mentioned above, the admissions committee may not be able to review your application with the others if it is late. I would recommend starting to gather your GRE scores and application letters a year before you intend to apply to schools. You may not like your initial GRE score and may want to retake it, an option you will not have if you take the exam a month before the applications are due.

Keep track of your proposal's status. It isn't unheard of for an application to slip behind an administrator's desk, never to be seen again. Anything can happen in the great hurried bureaucracy of a university. It is your responsibility to make sure that your application was received and is being reviewed. Let your potential advisors know that you mailed in your application. It would be a good idea to also send a copy of your application directly to your potential advisor.

Graduate Record Exam (GRE)

This standard requirement is an unfortunate one. My personal thoughts are that a student's grade on this exam is a poor reflection of how that student will do in graduate school. I've seen students with near perfect scores drop out of graduate school, whereas others who had laughably low scores go on to outstanding careers. When I say laughable, I mean "your verbal score is so low, perhaps we should require you to take the TOEFL (Test of English as a Foreign Language) instead" kind of laughable. So what is the point of taking the GRE? Well, it is the only standardized portion of the graduate application and the only quantitative portion. For this reason, scholarships are often distributed to students with the highest GRE scores. The GRE should not be something that you blow off. It should be something that you study for and see as an opportunity, as you would perhaps an IQ test. There are many administrators and faculty who put great importance on this exam. The verbal portion of the exam in particular is one on which admissions committees focus. It is generally thought to be the portion of the test that students can't really study for; at least it is the section on which you can't greatly improve your score by "cramming." This verbal portion of the exam might tell a professor how much time he could end up spending on editing rough drafts of an applicant's manuscripts. In that way, a poor grade on the GRE may be a deal breaker.

Study for the GRE in the same way you would for any standardized test: Take a lot of practice tests. Become familiar with the kinds of questions asked and your own strengths and weaknesses. In multiple-choice sections in which you are penalized for wrong answers, be sure to learn which parts of the exam you most often guess correctly and incorrectly. Be sure to take more chances guessing on the real exam on sections you generally guess correctly, while skipping those where you generally guess wrong. Make a note on your practice exams on questions that you guess and see how your grade would change overall. If you take practice exams, be sure to take recent versions; the GRE is a dynamic exam that is often changing. Public libraries are a great source for GRE prep books with practice exams, but it might be worth investing in a new one that someone hasn't marked up and that isn't outdated.

It is worth looking up tips on how the GRE exam is scored and how the adaptive exam changes based on the student's performance. How the exam is administered changes almost yearly. I recommend reading the Wikipedia page about the GRE (http://en.wikipedia.org/wiki/Graduate_Record_Examination) to learn more about how the exam is scored and to get tips on how to take the actual exam. For instance, if the exam is taken on computer, your final score for a section may be heavily biased by how well you do on the early questions relative to later questions. In this instance, it is worth taking your time early on, working to get the right answer in the first set of questions in each multiple-choice section, while perhaps speeding up the pace with later questions.

GRE scores are important in helping you obtain scholarships, so don't think they don't matter. A scholarship can insure that you don't have to teach to make a living as a graduate student, so you can focus on research and finish early. Some schools make you take GRE subject tests in addition to the regular exam. There are subject tests for biology, biochemistry, cell and molecular biology, chemistry, computer science, math, physics, psychology, and English literature. If you are applying to graduate school in any of these subjects, make sure you know what the programs require. If you are a physics major who wants to go to graduate school for psychology, you might consider taking the GRE in psychology even if it isn't required, to show the admissions committee that you have a strong enough background to succeed in their program. If you do poorly, you won't have to send in the score unless it is required.

Write a CV

If you don't have a Curriculum Vitae yet, start writing one. (Also, see the example CVs in Appendices 1 and 2 and the tips provided in Chapter 1.) You should have a personal CV that lists every major accomplishment, every conference, every experiment, and every hobby you have ever had. From the "Everything CV," you can fine-tune your actual application CVs, according to the program to which you are applying. For a graduate school

CV, you will want to put your name and contact information at the top, followed by any publications, research experience, conference presentations, grants, etc. (see example in Appendix 2). If you don't have any of those top four items, then you will have a hard time getting into graduate school. One good way to write your CV is to look at the CVs of the people you want to work with and their students (many of these can be found online). Use as many of the categories on their CVs in your own, but don't leave empty sections. Also include things like "Research Interests," where you can state some of the directions you would like your research to take. Keep these aspiration portions short (4 or 6 sentences); you don't want to emphasize your lack of experience.

Reference Letters

This portion of your graduate statements is largely out of your control, but if your references are negative, you will not get into a decent graduate school (see section on recommendation letters in Chapter 1). These letters are essentially character references. Most of what the admissions committee will do with these is skim them, looking for something negative. A negative comment like "he has a hard time learning new things" or "she has a hard time working with others" is a real blow to a candidate's chances. More seasoned researchers will never write anything negative in a student's recommendation letter, unless they really want to kill a student's chances. These researchers could be using their chance to a write a "recommendation" letter for a poor student as a warning to others.

Work hard to get letters from people you know; this is why you volunteered all those days and got to know your professors.

Get a Website/Cards

In the age of Facebook and Google+, etc., you want to have a professional website that you can have as a place to post your CV and pictures of your work and experiences. This is an easy way for potential advisors to learn about you and your work. Having a business card also isn't a bad idea. I don't recommend leaving cards lying around on a table like a real-estate agent, but handing these out to people can help expand your reach in the community, particularly at scientific meetings and when visiting a department. These also can help build collaborations. If you have a name that is hard to spell or pronounce (like I do), giving someone a business card with your website that has a link to your CV is an excellent way to make yourself memorable. Don't just pass out the business card either; before you pass it to the person, write a little note on it, such as "talked about working together on Panama project." That way, when the person empties his wallet of cards from several new acquaintances, yours won't end up in the trash with the other "who-was-this-again?" cards.

Be careful what you post on social networks like Facebook. Although we hate to admit it, a student's personal rants on politics, religion, or other causes may sway a decision either consciously or subconsciously. If you are going to post pictures from your drunken weekend in Cabo, then at least have the sense to set your privacy settings to restrict public views.

Recruitment Weekend

If you are lucky, and you are good, you will be invited to visit some of the institutions to which you applied. Some institutions bring in a student on their own, but increasingly popular is having a recruitment weekend for potential students. These recruitment weekends are typically fairly low-key events, during which prospective students are taken to see various sites around campus and the city, to see various labs, and to meet with faculty and students in mutual evaluation. These one-on-one meetings are very important, but so are the social meetings. (Don't be stupid, e.g., don't get drunk.) These recruitment weekends typically take place in February or March for a recruiting class for the fall semester.

The fall semester is typically when most students enter graduate school, and their applications are usually due at the end of the calendar year. This leaves a couple of months for the admissions committee to look over the applications and to narrow their choices down to the cream-of-the-crop plus some "almost-cream-of-the-croppers." The recruitment class is typically larger than the number of students the department is willing to accept. Depending on the school, some portion of the recruits will head elsewhere, and some portion will not make the cut. If a department is taking between 10 and 12 students, they might invite 15–20 students for a recruitment weekend. Typically, 70% of those invited are students that the department really wants, and unless something unforeseen happens, they are essentially guaranteed acceptance letters. This 70% are those who the department is really recruiting and trying hardest to impress. The other 30% are being examined to see whether they will make the cut. Many of the bottom 30% will also eventually get acceptance letters because of the attrition among the top tier (caused by these top students choosing to go to other schools). The very bottom 10% of the recruiting class are the ones that aren't quite making the cut but that can prove themselves as worthy during the recruitment weekend. No matter where you actually stand, you should prepare and conduct yourself as if you might be on the chopping block in that bottom 10%. Make no mistake: Everybody who is a recruit is still just a candidate and anyone can be cut. Depending on how many international students are accepted into the program (these students typically aren't invited to recruitment weekends because of the cost of flying them in), the bottom 10% of the invited prospects could actually be the bottom 50% of the overall recruits. There may even be other people recruited to the same lab you want to join. Typically, PIs only take one student at a time, but junior faculty (pre-tenure) in particular may accept multiple recruits. Also, PIs with a good track record with students often are allowed multiple students in a single year. However, if you are on a recruitment visit and there is someone else applying to the same lab, you might be in competition for a single position. If this is the case, just be yourself, work hard at promoting your good side and experience, and conduct yourself properly. Under no circumstance, should you try to undermine the other recruit. First of all, it is wrong. Second of all, that person may end up as your labmate.

From the recruit's perspective, the visit should be a time for you get to feel out your new potential home for the next few years. You should ask questions about stipends, housing, and internal grants (grants available through the department and university). And, most importantly, this gives you the chance to check out potential advisors. If your advisor is aloof and not very attentive during your recruitment, do you think he will be hands-on during your graduate career? Talk to the other current graduate students, check out the facilities, and learn how many classes you need to take and how many classes you need to teach. If you have more than one offer, you should weigh your choices carefully. If one place has the advisor you want but you will need to take three years of classes that you aren't interested in, you may want to choose another option. Perhaps the better option is the school that lets you jump straight into research with your second favorite advisor. My recommendation would be to choose the advisor that fits you best with all other factors carrying less weight. Some students make the mistake of choosing the university with the biggest brand recognition. This is great for an undergraduate degree, but it doesn't work in graduate school. You are the brand, and your output is all that will matter. The ideal situation is one where you find the advisor you want in a competitive, well-known program where you can focus on research and getting your degree.

Summary

- If you have the academic background to skip a master's degree, skip it. No need to add extra steps to achieving your ultimate goal.
- If you need to get a master's because you lack the experience to enter a Ph.D. program, consider programs without doctoral programs in which master's students often excel. Schools with great Ph.D. programs don't always have great opportunities for master's students.
- If you have a GPA higher than 3.4, better than average GRE scores, some graduate classwork, and solid research experience (with a publication or two), you should be in good shape as an applicant to almost any graduate school.

- The most important criteria by which you will be judged by an admissions committee is your fit with a potential academic advisor.
- Seek out advisors who will be good fits based on your shared research interests; read the publications of these potential advisors and contact them in a professional manner.
- Get reference letters from established people in your field and from professors with whom you have worked. Letters from PIs are better than from graduate students and lecturers.
- Statement letters with concrete goals explained as if in a research proposal are much better than letters full of lofty claims and wishes.
- Get started on preparing your graduate application a year before you intend to apply so that you can retake GREs and gather additional reference letters if needed.

3 Graduate School

The Hard Truth About Grad School

The best way to look at a graduate diploma is as a hall pass that enables you to do the kind of independent work you've always wanted to do. Too many students see earning a Ph.D. as reaching the top of the mountain: It ain't. Once you've obtained your degree, few people will care what your dissertation (thesis) was about; they will only want to know what products resulted from your time as a graduate student. They will want to see concrete products, such as patents or publications. The dissertation itself is basically just an unpublished treatise. Your dissertation will go on a shelf in your old university, and few people will ever look at it again. In fact, the most viewed dissertations are the ones that were never "published";

A Guide to Academia: Getting into and Surviving Grad School, Postdocs, and a Research Job,
First Edition. Prosanta Chakrabarty.
© 2012 John Wiley & Sons, Inc. Published 2012 by John Wiley & Sons, Inc.

i.e., the author never published the results from it in scientific journals, likely because he left academia. An unpublished thesis is like a carcass left out to rot: academic scavengers will take those results and turn them into something useful for their own publications. Don't let that happen to your work. Remember that you're getting the Ph.D. or master's degree so that you can move on to the next thing that you want to do, not just for the sake of being called a doctor (I hope).

Many students brazenly enter graduate school and think it will be relatively easy: It won't be. It won't be a 4-year research project, and it certainly will not be an extension of your bachelor's degree. Instead of trying to do well in classes, you will be asked to think independently and come up with a unique project that will lead to writing a book (i.e., your dissertation). You will learn that being smart will not be enough. You need to work really hard and you will need a thick skin to succeed. Advisors and reviewers will put down your best work, and you will forced to start over from scratch. Those that learn from these hard lessons move on; those that wallow in their own misery are quickly culled from the herd.

Whether you plan to use your degree to get a faculty position, research scientist job, liberal arts position, or nonacademic job, you will need to compete with research-minded students who want to be like their advisors (i.e., professors and independent researchers running a lab). Those professors will not train you differently because you have different goals than they did when they were students, and their expectations of you will not change because of those differing goals. Their job is to train you to get a Ph.D.; your job is to earn that degree by creating and carrying out a unique scientific project that you will write up as your thesis.

Learning your way around. Your closest mentors in graduate school are often not the faculty, or even your own major advisor (the scientist that accepted you into his lab). More often than not, your most important mentors will be your fellow graduate students. Every academic researcher hopes to have a student who works independently (i.e., a student who figures out things more or less on his own and who contributes positively to the lab). Your major advisor should be there for you when you need him (to help you learn to write scientifically, for writing recommendation letters, helping you apply for grants, and generally showing you the path to getting your degree), but often the people from whom you actually learn the tools you need to obtain a graduate degree are the later-stage graduate students and postdoctoral fellows in your lab or department. These more seasoned folks have made it to the next step and are where you want to be in the near future. They have recently gained tools that they are often willing to teach to the younger crop of students. They have the fresh experience that you need to gain. Rarely will they volunteer this vital information to you; you have to seek it out. The best way to do this is by collaborating with them and learning the skills they have in a hands-on manner. Whether it's how to use a particular tool in the lab, learning which grants to apply for, or figuring out how to get your advisor to listen to you, the people who know the best are the ones who have been there longer than you have. The sooner you learn from them, the better off you will be, so stick your hands in as many pots as you can by collaborating with your labmates. If you work hard, you will be included in other projects and hopefully can co-author publications stemming from them. Don't bite off more than you can chew, though. If you build a reputation as someone who doesn't deliver what they promise, you will be cut off from future projects. Despite this caveat, I would still suggest that you say "yes" to almost every collaboration you are offered. You will find the time to get it done if you work hard, and that will result in publications and learning the tools you will need for independent research. The more you put on your plate, the more products you will produce, and the more impressive your resume will be as a result.

The most important thing you will learn in graduate school—and, for many, this is the hardest to learn—is how to write scientific papers. This is where your major advisor will probably be the most help, being that he or she will be the most seasoned author in the lab. Learning the art of scientific communication is the key to a great career. You can discover how to use hydrolysis to power an airplane, but unless you know how to write a proper paper explaining your results, no one will ever know. There are plenty of students who are smart, energetic, and accomplish many things during graduate school, but there is only one scientific currency: the peer-reviewed publication. You can work on dozens of projects with interesting results, but unless they are written up and accepted by peer review (the process by which established scientists read and make comments on your submitted manuscript), they are not worth much. Go into graduate school with the hope of publishing every project of which you are a part, and always think of the publications you can write based on work done in a class or when working with a collaborator. Don't let others take away your chance to write up projects in which you designed the initial concept and carried out much of the work.

The only way you will learn how to write scientifically is by writing and writing and writing: practice, practice, practice! You should set a goal of trying to write two pages a day. This may sound impossible, but doing your best to achieve this goal by writing up the methods of analyses you've just started, or the introduction of an idea you had for a paper, is a great way to not only get practice writing but to have the majority of your work published. Lots of people have completed projects that have never been written up; by forcing yourself to write up each step of a project as it's done, you will end up with a nearly complete paper once the project is completed.

In the end, the criteria that will help you graduate and get a job is your publication record. If you have a five-chapter dissertation and three of them are published, your committee will likely not even read the published chapters because they have already passed peer review. Your publications will become your dissertation chapters. It is in your best interest to write up your projects as you go, adding layers to your dissertation as you steam roll toward defending your thesis and getting your degree.

When you have completed your thesis you will be the world's expert on the particular subject of your dissertation. Your former advisor will become your colleague, and you will join the postdoc and job markets.

Graduate school is a wonderful and formative time. You will never have this kind of time to study every angle of a subject again. In graduate school, your motto should be "I'll do everything." Do all the projects, collaborative work, volunteer work, and anything else you like; you won't regret it. I was once told, "if you want something done, give it to a busy person"; be that busy person, eat it all up. You'll regret missed opportunities more than the projects that didn't pan out. At the same time, you don't want to spend 15 years getting a Ph.D. Your goal should be an efficient, well thought-out project that will help you reach your career goals. Whether you want to go on to be a professor at a research I university (a top-tier research school), teach at a small liberal arts college, or get a nonacademic position, you will need to accomplish the goals set out by the graduate faculty and compete with students who want to go on to have an academic job. A good way to know where you stand is by meeting with your Ph.D. evaluation committee once a year. This committee, made up of your advisor and several other faculty, is there to evaluate you, and that is just what they will do. Even if they are not required to meet with you every year, it is probably a good idea to ask them to do so. This annual meeting will let you discover whether they think you are progressing at a good pace and it will also keep you from being surprised by something that might have been festering for a while. You don't want to find out during your thesis defense that a committee member was really upset that you didn't finish that collaboration with his postdoc. It's always best to be proactive about your graduate career; your advisors will take notice if you go the extra mile to stay on top of things. Remember that you are ultimately the one in charge of your career.

Seminars

As a graduate student, one of the best places to showcase yourself is during seminars, and not necessarily only during ones that you are giving. If you ask probing, interesting questions during a seminar, you can distinguish yourself from the silent masses. If there was something the speaker said that you didn't understand, chances are that few others understood either and they'll appreciate you asking. Be careful, as you can distinguish yourself in a negative way with an overly critical comment or a "gotcha" question aimed at a seminar speaker, but a good question can earn you kudos. Good questions often deal with methods: some variant of "how did that analysis work again?" or "you said [this] but does that method really test [that]?" are usually worth asking in nearly every talk.

I recommend going to as many seminars as you can, whether they are in your immediate interest area or not. Try to ask a question at each seminar, or at least write out a question or two during the talk that you could ask. You will find that the quality of questions and your confidence about them will increase with practice. I find that bringing a notebook and a pencil to a seminar allows me to take notes and formulate a proper question. I have several notebooks that hold notes on nearly every seminar I have ever attended, and I sometimes refer back to these notes when researching new methods or techniques. Having a notebook handy even during a very dull seminar (and you will go to many), gives you an opportunity to think about other projects or ideas that pop into your head. If you tend to fall asleep during seminars (which is the worst thing you can do), writing in the notebook as if you are in a class will keep you alert and engaged. Even during the dullest seminars, I've come up with good ideas for my own work. At the very least writing in a book during a seminar makes you look studious, using a smart phone, on the other hand, will make it look like you are checking your e-mail or sport's scores.

Giving a seminar, as opposed to just watching one, can be a stressful event. When you give a seminar, practice as much as you can beforehand. Practice until you are bored, and then practice some more. Don't worry about your presentation becoming stale; your energy level will go up again during the real thing due to nervousness or excitement. Try to focus on the major point of every slide instead of memorizing a script. Definitely don't just read off of a script when giving a seminar; few things are as dull as watching someone else read. You can have a script with you to help you remember some key points, but don't stare at the pages and read the entire thing. Making eye contact with your audience and using your presentation to direct them will make for a much more engaging talk. It's not just what you say, it's how you say it: Sell it!

As you begin to practice, print your slides and write the major points of what you would like to say on each slide; as you practice, make sure you hit those points. Remember that you are trying to convince your audience of something. You are trying to educate them on the points you are making. Imagine yourself as an audience member: What would it take to convince you that the conclusions are correct? Try to avoid acronyms and jargon that your audience will have a hard time remembering. Try to speak to everyone in the room, keeping in mind those with the least experience, particularly when explaining major points. Keep your speech simple and speak as you normally would, rather than in a lofty pedantic manner. You are trying to explain a project that you've done; you're not there to show off your vocabulary. The most enjoyable talks are the ones that can be understood: Don't let trying to sound like an expert cause you to lose your audience.

There are six categories that you should pay close attention to when making and presenting a scientific talk. Make a scoring sheet with the criteria listed below with a score of 1 to 10. As you practice, evaluate yourself, so that you can see where your weaknesses lie. Judges at conferences use a very similar scoring sheet for evaluating student talks.

(1) **Hypothesis and objectives** *(score 10 if clearly stated; 1 if not explicit)*—This should be what your introduction is about; your intro should not be just background. State up front what you are studying and why. In big, bold text state your major hypothesis. Remember that a hypothesis is stated as an answer to something (e.g., "Gravity causes things to fall"), and you are testing that answer in order to falsify or confirm it.

(2) **Methods** *(score 10 if sound and cutting edge; 1 if only briefly explained)*—Clearly state what you did to test your hypothesis. Students often skimp on explaining this to save time or because they think it is boring. Never do that! It's just plain bad science. The Methods section is one of the most important parts of your talk. People want to know, "Why did you use this program instead of that new thing?" or "Why did you take this particular approach to testing your hypothesis?" If you can't explain your methods backwards and forwards, you won't be able to convince anyone that you understand your own results or that your tests were adequate.

(3) **Results and Discussion** *(score 10 if justified and notable; 1 if limited or overreaching)*—Your results should be the facts about what your analysis found, and the discussion is an interpretation of those results. An example of a result could be the following: "The null hypothesis is falsified." A discussion might start with, "Falsification means gravitation theory has to be reinterpreted in a new way." Students often make the mistake of extrapolating their results into a grandiose discussion. Know the limits of your data and your tests. Be realistic in interpreting the significance of your research. It's great to be excited about your results, and okay to include something like, "I find it very interesting . . ." or "I was surprised and excited to get this result." It is unlikely that you've solved a problem so conclusively that it never needs to be examined again, so end your talk with a brief mention of what future work you plan to do to further refine your hypothesis. If you dwell too long on future work, you might weaken the strength of your current tests, so don't harp on the future too much.

(4) **Delivery** *(score 10 if calm and collected, good eye-contact and gave impressive answers to questions; 1 if no time is left for questions)*—Use your time effectively. If you are giving a 15-minute presentation at a conference, you should have a 3- to 5-minute intro and another 5 minutes for methods and results. The discussion should be succinct and hard-hitting, so no more than 3–5 minutes here either. Leave at least 2 minutes for questions. If you are being judged for an award at a conference, you are usually disqualified if you don't leave time for questions. One way to save time is to only read a

few important names on your acknowledgments slide. You can also leave that slide on the screen during your question period without reading it. Reading 35 names out loud with an explanation of how each person helped you is not a good use of your time. I'm always surprised how many students run out the clock by doing this. It gives the audience the impression that the speaker is trying to avoid questions.

(5) **Visual Presentation** *(score 10 if appealing and creative; 1 if text is undecipherable and videos didn't function or were not pertinent)*—Your slides should look professional and polished. Try to keep the background and font the same on each slide. Don't use too many colors or animations that are distractions. On the other hand, a white background with black text and no images is not visually appealing to anyone. Relevant images are not only nice visual aids, but they also help with the esthetics, as do relevant animations and videos (but make sure these work on the conference computer and projector!). Also avoid patterns that the color blind can't see, especially red and green. Blue backgrounds are great, but if mixed with red or orange fonts there won't be enough contrast for the audience to read your slides well. Always side on having more contrast in color between the background and the text with the lighter color being the background. This way, if the projector is not very powerful or if the overhead lights don't turn off, the audience will still be able to see your text without having to squint.

(6) **Quality of research** *(score 10 if creative, rigorous, and significant; 1 if ordinary and insufficient)*—In this somewhat subjective category, judges evaluate the (a) creativity, (b) scientific rigor, and (c) significance of your research. The best way to get good scores here is by (a) highlighting the interesting way you tacked the scientific question (i.e., creativity), (b) explaining how you kept working on the analysis beyond the typical evaluation (i.e., scientific rigor), and (c) explaining how you interpreted your data to be important in the larger scheme of things (i.e., significance).

For any presentation, humor is a great way to cut the dry, stuffy atmosphere but be careful not to over do it. Starting a talk with a joke usually doesn't pan out, especially if you are nervous; just let your natural sense of humor do the work if something witty comes to mind and the mood is right. If the witticism comes across as being spontaneous (even if it isn't), it usually works better than a prepared joke. If you end up saying something that comes out awkwardly, don't sweat it too much; the best thing to do is move on. The more you practice, the lower the chance of something awkward happening.

It's okay to be nervous, but don't psych yourself out. Having attended many presentations at this point, you know what makes for a great talk. As a practice exercise, imagine how a speaker you admire would present your talk? You don't need to mimic their mannerisms, but it might help you think outside of yourself and put your talk in a different perspective. You want to be careful, too; just because professors can act aloof and completely nonchalant during a professional presentation, doesn't mean that you, as a student, can get away with pulling off that same approach.

Long introductions can cause you to lose your audience, so get to the point as soon as you can. Just as with writing, your first few sentences should directly inform the audience about your topic. If you are giving a 45-minute presentation, your data should show up in the first 10 minutes. Even if you are introducing a complicated study, the background should be explained quickly and in simple terms. People want to know what the "big picture" is from your work. Just as in a scientific paper, start with broad ideas and narrow them down as you go, ending with a broad conclusion. Know your audience. If you are talking at a genetics conference, you don't have to explain what DNA is. If your talk covers multiple projects (something you should only be doing in a longer talk), your introduction should explain the common thread that unites them; explain each project in a separate section with its own conclusion slides, and end the talk with an overall summary conclusion slide that again brings them all together. This will help the audience understand the value of each component within your overall theme. After you get through a long series of result slides, interject comments like, "Now this is an interesting result because…" or "This is very exciting to me because…" or "Well, what does this all mean then…" These little interjections of vernacular speech help lighten the mood and draw the audience's attention back to you as a speaker. In scientific papers, you want to use efficient scientific language, but when you are speaking to your peers, you want to talk directly to them as you normally would (perhaps with a bit more flair). It is okay to use everyday speech to get your point across. Break down the barriers between you and the audience; you are not reading a scientific paper out loud, you are explaining one. If you really know what you are talking about you will be able to explain what you are doing so that everyone in the room can understand.

If you get very nervous about your talk, try focusing on the first few slides when you practice. If you can get through the introduction, the rest should come naturally: this is your work you are presenting after all, no one knows it better than you. Just before your talk, try to stay loose. Rather than sitting alone getting nervous, talk to people you know who will calm you down. Talk about baseball and beer or whatever will loosen you up. Try to focus on the first thing that you will say, especially if you're afraid that you'll draw a blank. You may want to write down the first sentences—"Thank you all for coming. I'll be talking today about sex changes in crickets"—and bring them to the podium with you. If you have a quiet voice, try to speak to someone in the back of the room or use a microphone.

If you need an icebreaker at the beginning of your talk, particularly for longer nonconference talks, asking the audience a direct question can help get everyone settled in. Asking the audience a question or two can also help you gauge how much background knowledge they have. For example, if you are talking about global warming ask "Does anyone know what the Earth's average temperature is?" and then solicit a few guesses. This interaction will calm you down and loosen up the atmosphere. Again, know your audience. This strategy works for longer presentations, but you often don't have much time for audience participation in a 15-minute conference talk.

Try to preview your presentation in the same room where you will give it. If you can, give a practice talk there with an audience of your friends and advisors. If you can't get an audience, at least visualize how you will give the talk; decide if you'll stand at the podium or if you will walk around. Most importantly, check that your talk will look to your audience the way it looks on your computer screen. Practice talking to the audience instead of talking to the screen behind you; eye contact will help you engage the audience. Alternatively, you can look at the back wall of the room, which will also help you project your voice. Practice using the laser pointer; use it sparingly and only when needed. Frivolous use of the laser pointer is very distracting. Drawing 3,000 circles around the entirety of your slide doesn't point anything out except that you are nervous. Repress your inner John Madden and use the pointer to point something out with a nice steady dot, not with frantic circling.

Aside from the introduction, the thing that most people get nervous about is the question period. Try to anticipate the kinds of questions you will get and add slides at the end of the talk that you can bring up if you need help answering tough questions. Put these slides after your final acknowledgments slide and only use them if you need to in the question period. If you are worried that someone will ask you about a program or method you can't explain very well verbally, make an extra slide with a detailed explanation and summary. Again, don't worry too much; you know this material better than anyone in the room, and you'll do fine. If you practice in front of a small audience of peers, the questions they ask are often the same ones that the real audience will end up asking. So practice your talk with an audience and also practice answering the expected questions. You'll look like a pro when it comes to the real thing.

Try to avoid too much text on your slides. As you practice and get more comfortable with the topic, reduce the number of words on each slide so that people are listening and watching you instead of reading and focusing on the words on the screen. Try not to linger too long on any one slide, particularly the first few slides. In the YouTube/iPhone age, people want to see things moving along; staring at a static image will only lead to daydreaming. Instead of simple black and white slides full of text, use colors and images that can engage people. If only text appears on the screen, people will tend to read that text and not listen to you. Keep your slides simple; too many animations and unnecessary text can be a terrible distraction, but a few slide transitions and informative videos can help keep the audience engaged. An interjection such as, "I hope this is clear, does anyone have any questions at this point?" can help loosen up the crowd if you sense their attention is fading. Try to engage your audience periodically throughout the talk. Instead of standing there lecturing for 45 minutes, break up the lecture with a question to the audience, a video clip showing some part of your project, or some other element that can change the pace a bit.

The worst thing you can do is talk for too long and go beyond your allotted time. You can be talking about the most interesting thing in the world, but if you keep people in their seats longer than they anticipated, you will be upsetting them. Always remember to leave time for people to get settled at the beginning and time for questions at the end. At a university, when classes are in session, typically a talk scheduled for 1 hour between 12:00 and 1:00 will start either 10 minutes after the hour (12:10) or end 10 minutes early (12:50) to allow people to get between classes. Assuming the first option, if you talk for 45 minutes, it will be 12:55 when you finish, leaving five minutes for questions. Try to time your talk every time you practice it. Be aware of the time and your time limit during your talk by bringing a watch or stopwatch or by using the clock function on your computer. Going over your allotted time is a no-no, and people quickly lose interest with each passing extra minute. Ideally, this time (i.e., the end of your talk) should be when you are going over your discussion and major conclusions, so it obviously should not be the time that you want your audience's focus to be fading.

When answering questions, try to answer them as clearly as you can. You can interject some additional points, but don't spend five minutes answering one question. Try to answer the question, and then move on to the next one. The question period is when you should be (or at least appear) the most relaxed; step away from the podium and engage each questioner directly.

When you end your talk, simply say "thank you" not "and with that I'll take any questions." If you end with the latter, people won't clap because you've started the question period; ending with "thank you" means you will get some applause and a breather during which you can take a sip of water. (Always remember to bring water!) After the question period, be prepared to stick around. If this is a talk at a conference, this means staying until the entire session is over. There are usually more questions than can be answered in your allotted time, and this post-session period is also when potential collaborators will come to you to set up a meeting.

I generally recommend giving a seminar rather than a poster at meetings, but this is largely dependent on the target audience. If you want to reach a lot of people in a short period of time, nothing beats a seminar. If you want to reach a few people (<15), and really engage them, a poster might be better. If you go with a poster, don't just stand next to it waiting for someone to engage you; you need to be the initiator. Few people actually want to read all the text in your poster; your poster is a hook for you to describe your research. If someone walks by and gives your poster more than a passing glance say, "Hi, can I give you a quick overview of my work?" Then, using the figures on the poster as a guide, give them a two- to three-minute presentation. "I was interested in knowing why [pointing to Title or Intro figure] ... so I [did this, pointing to Methods] ... then I found [pointing to a figure in the Results] ... and I realized this was important because [give Conclusions]." Let them ask questions, and draw them in. If you see someone you admire professionally, this is your time to show off. You might even want to have a one-page handout that summarizes your poster to give out once you complete your presentation; this is also a polite way to say goodbye and to get stragglers to move along. Try to engage as many people as you can; you never know who you might impress.

Tool Up Early

The beginning of graduate school is a strange time. You are basically set free to blaze your own trail or to carry out experiments related to your advisor's projects. Whatever the case, you should make it a point to meet with your main advisor at least once a week for a progress report, even if it is an informal one. This will help you set goals and judge your progress and also help your advisor decide what you need to learn.

Your first year of graduate school is often about learning the techniques you will need to complete your dissertation. You will likely learn more from senior graduate students and postdocs than from the faculty, including your own advisors. Many seasoned advisors have become more administrators of research than hands-on researchers themselves. Students are the ones coming in reinvigorating the lab with new techniques and gathering the raw data. These techniques can sometimes be learned from people in your own lab, although not always. You may find that volunteering in another lab to do some work in order to learn a technique will benefit you in the long run. If you need to learn how to do a particular new technique, you may find that a faculty member other than your advisor may be willing to have you work with them and get some projects done. They will get a free part-time researcher, and you will gain valuable research experience that you can apply to your own lab work and, potentially, a publication.

Tooling up (learning skills and techniques) early in graduate school is also a good strategy because later in your graduate career you will want to get results quickly and be analyzing them. You don't want to spend your time learning too many new tools later in your graduate studies; after a certain point, you will need to have measurable progress toward your degree, and if you are unable to show this progress because you are still learning new techniques, then your committee may wonder where your time has been spent. A lot of students spend more time preparing for qualifying exams than working toward their dissertation by carrying out and writing up projects. If you have publications and grants by the time you go up for these qualifying exams, those accomplishments will be more impressive to the committee than if you can answer every theoretical question they ask you. They will be much more lenient to a student who has already shown they can make it as a scientist than to a student who just knows how to talk like one.

The Notebook and the Pencil

One of the pet peeves many advisors share is having a student arrive to a meeting to learn about a new project with nothing to take notes with. Always bring a pen and a notebook to show that you will pay attention to the long list of instructions that you have to carry out. Coming with nothing in hand is telling your advisor that you did not show up today to be helpful or to learn for your own sake. Arriving with nothing is telling the professor that you want them to provide everything for you and that you will likely need most, if not all, of the information repeated later. When meeting a faculty advisor or even another graduate student, come prepared. The same notebook that you use for seminars will come in handy in these situations. A bunch of loose papers can be easily lost and become disorganized, so a nice solid notebook is preferable.

Different Kinds of Advisors

If you have a very hands-off advisor, you may want to stay close to senior students and postdocs in the lab to learn techniques, but keep in mind that you will need that advisor to sign papers, write letters of recommendation, and generally help you take the right path in completing your degree. You should always have your advisor read and edit your manuscripts and grant applications, as well as watch and judge your practice presentations. Seasoned researchers such as your advisor are the kinds of people who will be judging you throughout your career. If you have a more hands-on advisor (perhaps a younger untenured professor), you may be under more pressure to perform and keep things running in the lab. Depending on your own personal preference, you may like a particular kind of advisor more than another; it is best to find out which you have before you are stuck. If you find yourself in a situation that isn't working out because you are not getting the kind of advising you need, you may want to transfer labs (or even schools). It might be to your advantage to make the first move if a divorce is inevitable. An advisor who has lost faith in his student could put that student's career in jeopardy.

Your primary advisor is the most important person in your graduate career; you need your advisor's support to become a candidate and continue in your studies and onto a real career. Be sure to develop a healthy working relationship with them, while in their tutelage.

Candidacy

Every graduate program has some way of filtering out students who aren't cutting the mustard. Typically, in the second or third year (earlier for a master's), there are some major candidacy-qualifying exams that you need to pass, and they aren't your run-of-the-mill, multiple-choice exams either. In many programs, you need to give a presentation to the department showing a research project that you have completed or that is in progress. This presentation usually takes place early in your second or third year. Other programs may ask for a research paper to be submitted to your qualifying-exam evaluations committee. This committee usually includes your graduate advisor and two to five other faculty members that together will ultimately decide your fate and whether you can advance to candidacy. Being a "candidate" typically means that you are a graduate student who has (1) completed all exams and coursework, (2) is in good standing, and (3) is securely on the path to getting an advanced graduate degree. If you fail this candidacy-qualifying exam, you may be asked to leave the program. At the very least, you will have to take the exam again. Being a candidate opens up new grant opportunities within your own program but also with NSF, NIH, and other agencies.

You should see these tests toward candidacy (sometimes called "Quals" for qualifiers, "Prelims" for preliminary exam, "Generals," or the like) as opportunities. Make the requirements from these tests into things that will be useful for you (i.e., dissertation projects). For example, if you are asked to write a review paper, then turn this into the intro/first chapter of your dissertation and a publication. If you need to present a proposal for your thesis, write it in the format of a NSF Doctoral Dissertation Improvement Grant or other grant proposal. At some universities, one of the qualifying exams is typically a written exam. If this exam is usually administered as a series of written questions, it likely won't be very useful to you. It can be to your benefit to ask if you could fulfill the requirements of this exam by writing a dissertation chapter instead. The rules about these exams are found in your graduate school handbook and are often written vaguely enough that you can ask for the exam to be administered in a way that will be useful to you and that will show the committee you are serious about advancing. You will need the approval of your committee to make changes to your exam, but they will generally agree as long as it is helping you toward your goal of completing your dissertation. Usually, if your advisor agrees first, he can sway the rest of the committee.

Almost universally, one of the tests for passing candidacy is some kind of sit-down meeting/grilling with your committee, sometimes called the "oral exam." These are always tough to prepare for, and almost no one performs well in these situations. However, these sessions are good practice for job interviews, except they last much longer and the people asking them are tougher because they know you better. These 2- to 5-hour-long sessions usually revolve around topics related to something you have written for another portion of the qualifying exam (e.g., your dissertation proposal, first chapter) so try to steer the committee's questions toward the topics you know best. Questions peripheral to the topics you know well are where you are most likely to flounder. You will be nervous, but try your best to listen to the questions being asked; sometimes repeating the question will help jolt something in your brain to help articulate an answer. If the answer doesn't come to mind right away, ask if you can return to that topic a little later; sometimes an answer to a different question will refresh your memory. Be as prepared as possible and try to learn facts about what types of questions were asked in these sessions from students that took the exam previously. If you have a good relationship with your committee members, you can even be so bold as to ask them ahead of time what they expect you to know and what areas of research they may inquire about. Typically, the entire cohort of students that started the graduate program together goes up for candidacy at around the same time. If everyone in that cohort has the same qualifying committee, then going first may be a good strategy. The professors will fine-tune their questions with each student, and the questions tend to get harder. On the other hand, each subsequent student will get to hear how the committee acted and what questions were asked. In the end, these sessions determine the gaps in your knowledge. They want to know if you really know what you are talking about and to see if you have sufficient enough of a background in your field that you won't need to take additional classes. If you don't pass the exam in your first attempt, you may be asked to take a class in a subject area that you didn't quite seem to grasp. You may even be asked to retake the exam. Even the best students sometimes do poorly on these sessions, but the worst students almost always do. You should be able to explain every detail of the things that you are expected to know the best; the subjects that you intend to cover in your dissertation should be topics you know backward and forward.

Be humble, and don't be argumentative during the exams. If you think you are being attacked, chill out. You are not being grilled because the committee doesn't like you; they are pushing you because everyone gets roughed up a little in these exams (it's a ritual hazing, like it or not). You have to let your guard down a bit and take the heat for a while. If you don't know the answer to a question, don't just start saying anything, or blabber on off-topic. If you really don't know, rub your chin inquisitively, think about it for a few seconds, and then say, "I'm sorry, I don't know." If the committee thinks it is a very important question, they may ask you in a different way and lead you to the answer. Bring some scratch paper to write down the questions if that helps jog your memory. It might be in your best interest to bring water and maybe even light snacks for you and your committee.

At no other time will the split between student and teacher be greater and more evident than during these oral exams. You might be best friends with your advisor and go drink beer with him every Friday (or every day), but during these exams, he has to put on his advisor cap and grill you just like the rest of the committee. Remember, they are also being evaluated, young or old; they need to show that they are still relevant or that they are training students properly. If their students are being evaluated, so are they; if you perform badly, that makes them look bad. If they have to explain why their students performed poorly, they might throw you under the bus to save their own hides.

After they are done grilling you, the committee will ask you to step outside while they discuss your performance. In your absence, they will decide if you pass or fail. Conversations like this are known to happen:

"Dave, how come your student never learned this stuff?"

"I don't know, I told him he would be asked exactly this sort of question. I'm really disappointed, I'm not sure if I want to support him as a student anymore."

How do you avoid this happening? Study, study, study; but, more importantly, prove that you are a worthy candidate by publishing and performing at the highest level before you even step into the room. Prepare as much as you can. Being scared is good; if you are scared that means you are not overconfident and cocky. The overconfident often end up with egg on their faces. After the committee makes their decision, they will call you back into the room so that you may learn your fate. Perhaps they'll shake your hand, tell you that you've passed, sign your candidacy papers, and invite you out for a drink. Perhaps you will pass conditionally, in which case they might say that, although you've passed, you will need to take a course to strengthen your knowledge on a particular subject. You will probably need to get an A or a B in that class to become a candidate. Perhaps when they lead you back into the room your advisor will tell you that the committee has decided not to pass you based on your performance on the exam. In other cases, it might be a few long weeks before you find out how you did on this exam as all the students are evaluated together. If you have a longer wait, just go about your business. If you think you did poorly, there is no use sulking; just keep pushing on. It's often the students who think they did the worst that come out okay; the overconfident students are the ones who should brace for a fall.

If you fail, learn something from your failure. Sometimes professors are trying to knock you down a peg if they think you are cocky. You might have to hang your head for a while, especially if the rest of your cohort passed. It is time to reflect, but it isn't time to give up. You could have failed for multiple reasons; usually, it is because you did not perform the way you were expected to on these exams, not because you are not wanted. If that is the case, just retake the exam having learned from your mistakes. If there is a three-part evaluation and you failed all three parts, the signal from your advisor may be that he wants you out of his lab. If that is the case, you need to have a frank discussion with your advisor. I would also recommend talking to the other members of the evaluation committee. It might be that your advisor was the one pushing you out the door, whereas the other committee members saw you in a more favorable light. In that situation, you will have to switch labs to survive.

Most of you won't fail, and most of you will become candidates. Congratulations, part two of your graduate career will now begin. Now you can set all of your efforts on writing and ultimately defending your thesis.

Working with Other Graduate Students

As I've mentioned, the people you will learn from most are your peers. These peers will follow you throughout your career, and you will be surprised to see whom among them you end up meeting again as faculty at other universities. One day, they may be reviewing your papers or even reviewing you. They may help you get, or lose, a job in the not too distant future.

Most first-year graduate students are pretty naïve, but this doesn't keep them from being cocky. This bad mix of characteristics can breed intense competition among a cohort of incoming graduate students. Competition is a good thing as long as it makes everyone better and no one is cheating or hiding things from anyone else. Students who sabotage others are eventually ostracized by their peers; students who help their peers are rewarded. If you hear of a new grant that has become available, don't keep it to yourself. Being a good citizen always pays off, even though it doesn't always seem like it at first. The faculty will take notice and more likely remember you more than those students who keep to themselves and don't volunteer. Sign up to be on social or academic committees that you are interested in. I would recommend running some of the social committees early on in your graduate career and transition to academic committees later. Social committees, in particular those in which you can run the weekly graduate student coffee/beer session, or set up a workshop, can really be something that people appreciate. Later in your graduate career, you might be asked to be a student member of the graduate admissions committee (the committee that picks the incoming graduate class) or executive committee (the committee in which the faculty meet to discuss policies). You can learn a lot from being on these faculty committees, and it is good experience for when you are a professor yourself. If you are going to volunteer to do something, don't complain about the work. Everyone knows how hard you are working and appreciates it, even if they don't say it. If you complain about volunteer work, you diminish your efforts in the eyes of others. Your volunteer efforts will likely be remembered when it comes time to hand out awards and internal grants, and most major funding agencies also like to see a strong list of synergistic activities in a student's CV. Recruiting on behalf of the university or judging talks at a national meeting are examples of activities that these agencies like to see. Some of the best kinds of synergistic activities are those in which a student volunteers to teach local high school or elementary school students. Many granting agencies want you to include "Broader Impact" statements in grant applications that explain how your project will benefit society. If you don't have a good history of synergistic activities, these "Broader Impact" statements will look like empty promises.

Try your best to be a personable student who doesn't step on anyone's toes. Selfishness and laziness can kill a person's chances of getting a good position in the future. A bad reputation will not only hurt you at your present school but in your next position, too; the academic world is much smaller than you might imagine. It is never too late to start redeeming yourself; it's all part of the maturation process. There is also no room for whiners in graduate school. Whether you are a great student or a mediocre one, you will receive your fair share of negative comments from reviewers and advisors. Those students who take these critiques in stride do a lot better than those who take them personally. Even when criticisms appear personal, they rarely are; usually they're just poorly written and rushed. Nevertheless, there are few pats on the back in science; you might get some awards and grants and jobs, but mostly everything else is a criticism. You need to develop a thick skin about negative criticism. Sometimes it will be undeserved, but often it is not. It may seem unfair sometimes, but no one is immune to it. Learn from it and improve yourself; that is the only way to move on.

Classes

Don't take any classes that the graduate program doesn't require or that you don't really need to take. You will learn most things that you need for your dissertation by doing hands-on work. This is very different from the heavy class load that engineering or medical students experience out of necessity. You didn't go to graduate school to take classes; you are there to do original research. At the same time, there are often some wonderful graduate-level classes taught by famous people or wonderful teachers (or both); you should audit these (attend but don't register) unless you absolutely need to take them. Classes take enormous amounts of time, and you should have the freedom to take away what you want from them without the obligation of exams and assignments. If you intend to teach a particular graduate-level class in the future, I would recommend that you take that class as a student and keep your notes. This way you will not have to prepare a course from scratch in the future.

Your GPA at the graduate level doesn't matter very much unless there are some failing grades, otherwise grades are often inflated; that's why a C is typically viewed as a failing graduate grade. Remember that if you are asked to take a class as part of passing candidacy, you might need a B to pass. Unless you are planning on transferring these graduate classes to a new institution, your grasp of the concepts presented matters more than the actual grades. However, if you are a master's student planning on applying to a Ph.D. program, you should make the most of your classes so that you don't have to take them again in the future. In general, if you are required to take a class, you should try your best to do well in it. The professor might be someone who talks to your advisor or who may judge your grant one day. If you slack off, it will reflect poorly on you.

Publications

Try to only work on projects that will lead to publications. Most of your publications will likely be centered on your thesis chapters, but when you apply for a job the committee will ask, "What else did you do aside from your thesis?" When starting out as a new graduate student, one of the best ways to start working toward increasing your publication list is by writing a review paper. Writing a review paper about the subject you are most interested in will help you accomplish a number of important goals: (1) You will gather the important literature for a subject that you will likely use for many years to come. (2) You can find the gaps in the literature and in the science of your topic, allowing you to fill that gap with your own future work. (3) Review papers are often highly cited and make for excellent publications. (4) A review paper can serve as an excellent introductory first chapter for your thesis. For all of these reasons, I often recommend to incoming students that they write a review paper with 100 or so references in it. As you are reading and gathering these papers, you should keep track of what you have read. One of the best ways is by using reference managers like EndNote to keep track of what you have read and to make notes. I also recommend adding a few cited sentences or paragraphs from each reference into a draft of your paper. This will help you organize your thoughts and pick out the facts that you think will be most important; it will also help you remember why each paper was pertinent.

As I mentioned before, try to write two pages a day. Having this goal will ensure that you have practice in the most difficult part of being a scientist—communicating your work to others. This goal will also ensure that you are continuously publishing. I was once told that you should always have a paper in prep (preparation), a paper in review (submitted and being reviewed by colleagues), and a paper in press (a paper that has passed the review process, is accepted, and is awaiting publication.) (Read the H-index section in Chapter 7 if you are not aware what that index is.)

In each year of your graduate education, you should submit an abstract to a scientific meeting, with the intention of turning that project into a publication after the meeting. Abstracts are typically due several months before a meeting. Submit an abstract for a project that you plan to have results for by the time you get to the meeting. You will be putting a lot of pressure on yourself to get the project done, but it will be worth it. Setting tough goals can get you motivated and keep you ahead of the curve. The point of doing this is to keep a steady pace of publishing. Now that you've submitted an abstract, you can write a preliminary introduction and methods section for a future paper. Once you've given your talk, you can write up the results and discussion. Signing up for talks will not only keep you pushing yourself to publish, but giving the talk itself will also help you organize your thoughts and see the big picture themes of your research.

After the conference, put the finishing touches on the paper and submit it as soon as you can (the next round of abstract submission deadlines are just around the corner). By the time the paper is in review, it will be time for the next meeting. In this way, you can have a nice cycle each year of (1) abstract submission, (2) carrying out experiments gathering data, (3) writing up the introduction and methods, (4) giving talks at meetings, and (5) writing up the discussion and results to finish up the paper/chapter. If you push yourself, you can accomplish more than you thought possible. If you take the easy route and let conferences and time pass by, you'll find yourself falling behind. It's always a good strategy to set yearly goals; by signing up for a talk at a conference, you are tying your goal to something concrete on a fixed date.

Some students labor over publications for months (or years) trying to get every aspect of the project perfect before writing up the paper and submitting. This is an inefficient approach, and I worry much more about my colleagues and friends who are perfectionists than I do about those who may publish before every detail is in place. The perfectionist only writes up something every so often and, although it may be wonderful, the results come out so infrequently that they are not competitive academics (publish or perish, as they say). Those who are quicker to publish often have more collaborators, too, which leads to even more publications. No one wants to have their publication held up by a slow collaborator. Remember that it can take months for a paper to be reviewed and more than a year between submission and actually seeing the paper in print (this wait can be the most frustrating part of science). You should have your study as complete as possible when you first submit, but some of the minor details of a publication can sometimes be hammered out while your paper is in the first round of reviews.

Getting your work published should be relatively easy if your results are straightforward and the writing is clear. The hardest part is always getting past reviewers. Reviewers can be completely wrong in their assessments, but typically they make good points (even if you don't want to hear them). The reviewer's comments, despite the extra work they are asking you to do, usually make your paper better. Publish with the knowledge that the managing editors of a journal get tons of papers and have exhausted their go-to reviewers long ago. Giving a list of two or three potential reviewers in your cover letter to the editor can help you immensely in getting your paper through smoothly. Make sure these reviewers are not just your labmates because the editor will be able to tell; however, it is okay if you choose colleagues who are knowledgeable on the subject. In fact, those are exactly the people you should recommend, especially if they are efficient reviewers (i.e., they can return the manuscript quickly). You can also ask the editor to keep your paper away from some potential reviewers who might have an unfairly critical view of your research. (Your research isn't that interesting if you haven't made a few enemies along the way.)

You should also have a strategy for sending your paper back to the editor after you receive the reviews. Sometimes the deadline for returning your reviewed paper is very short (1 month is not uncommon). This is so the journal can maintain a rapid turn-around rate so it looks good for people who want to publish quickly. If you need more time, ask the editor for it; they almost always oblige. If you rush your paper, it is more likely to be rejected, especially if you didn't respond to all the reviewers' comments in your revision. The thing you want to avoid the most is having your paper sent back out to be reviewed a second time. As an editor, I can tell you that papers that I get back with obvious grammatical and spelling errors that should have been fixed are the ones I am most inclined to send out for a second round of reviews. If it looks like all the changes I asked for were made and the paper looks like it is in good shape, then I am more inclined to send it to press instead of back out for review.

Before you return your paper to the editor, make sure the obvious things like spelling and grammar are fixed. Also make sure that any "Track Changes" are no longer visible. You want your paper to go back to the editor looking like the finished product.

You should also spend a good deal of time on your rebuttal letter. This is the letter in which you explain the changes you made in response to reviewers or why you disagree with a reviewer's comments. Sometimes reviewers have opposing views on the same item, and you can use one's comments against the other. Typically, the editor has weighed in on these, and you should take his comments the most seriously. Make it as easy as possible for the editor to determine what changes you've made. Write a rebuttal letter that has a short note at the beginning telling the editor of all of the major changes that were made, particularly those regarding his suggested changes. Then add each of the reviewer's original comments with your comments underneath for each point (with your text in a different font or color). This kind of point-by-point rebuttal saves the editor enormous time comparing documents, particularly if you state which line or section you added the changes to in the new document. Try to make all the changes requested by the reviewers, but if you don't agree with a suggestion, make sure you make a clear, concise, and convincing rebuttal.

Being a Teaching Assistant (TA)

Unless you are on some sort of amazing scholarship, you are generally expected to teach in order to have your tuition paid for and to collect a salary as a graduate student. You will likely have to teach a lab portion of a class and possibly reading groups and review sessions as well. Being a teaching assistant can be a harrowing experience that can take over your life if you let it, especially your first time around. If you have a choice, try to teach smaller classes that have few written assignments and essay exams. Teaching a big class where you have to grade lots of essays can be a nightmare, as anyone who has to read a sentence written by a typical college student can attest. The best kinds of classes are ones where you have the most control. I liked being a part of courses with multiple TAs, where I taught later in the week after more senior TAs who had taught that particular course before. In these classes, instead of doing outside prep time, I could sit in on the beginning of the earlier labs and I could act like a student absorbing the lesson so that I could repeat what I learned to my own sections. "Teaching is to learn twice," after all. This would save me and the senior TAs time, since we didn't need a separate session on our own to prep. Teaching later sessions than the senior TA's saves you prep time as you can reuse the handouts and sometimes even the illustrations left on the board from previous sessions (see Box 3.1). If you do lean heavily on the senior TA in this way, you should volunteer to help in some other aspect, such as writing exams or some other chore to keep the workload you put into the class somewhat even with the other TA's efforts.

Graduate students often waste a lot of time trying to be an A+ student in the class that they're teaching. The difference between a B student and an A+ student is probably 10 hours a week of study time. In most situations, you do not need to be a perfect student, but you should be a competent student (i.e., a high B student). This means that you are someone who knows how to get to an answer, even when it may not necessarily be on the tip of your tongue. Leave memorization to someone who actually has to take the exam for a grade. Knowing where to find the answer is just as good as knowing the answer when you are in the role of a teacher. If a student asks a question you don't know the answer to, you can turn around and ask if anyone in the class knows the answer or refer to lecture notes or the textbook. If you still can't find the answer, tell them that you will find out and get back to them. There is nothing wrong with saying "I don't know," as long as you find an answer in a timely manner. However, there are likely a few classes that you will TA in which you actually should be an A+ student. If the class is on a subject related to your research, you should know more than any undergraduate student in the class. No matter what, you should always ensure that the undergraduates are well prepared; that is, after all, your responsibility.

It can take a long time to prep the first time you teach a class as a TA, but it should take considerably less time in subsequent intervals. Although teaching a number of different courses does increase the number of preps you have to do, diversity looks good on a CV, particularly if you plan on teaching those courses later as a professor or lecturer. However, if your primary interest is research, you should aim to teach as few different courses as possible, and only those useful to you, to minimize the number of preps you have to do on your way to getting your degree. If you are interested in an undergraduate class, it is better to TA that class than to actually register to take it as a student; you will learn more as the TA, and you'll be exempt from the headache of having to actually take exams.

Although all TA assignments require a time commitment, you may want to avoid courses in which grades are based on essays and other long writing assignments. You often end up spending much of your time being an editor of grammar and spelling, whereas content, so frequently buried by poor writing, becomes difficult to judge. Another reason to avoid these classes is because every year you will still be stuck spending the same 20 hours per week grading assignments: the work doesn't get easier because the students (and thus their work) are different every year. If you are in this situation, it is probably worth the effort of switching to a new course where all the grades are based on multiple-choice exams or something else that takes relatively little time to grade. (Of course, if you like the class and the editorial work, more power to you; just don't expect it to be a time saver.)

Keep notes and all the course materials for each course you TA, and you will thank your past self if you have to teach the same class again. You may even teach the class as a faculty member at some other university five years down the road, and these notes will come in handy.

Tips on Managing Your Time

The famous evolutionary biologist E.O. Wilson once said, "Over the years I have been presumptuous enough to counsel new Ph.D.'s in biology as follows: If you choose an academic career you will need forty hours a week to perform teaching and administrative duties, another twenty hours on top of that to conduct respectable research, and still another twenty to accomplish really important research." I largely agree with that statement, but I've found that the amount of time I've physically spent at work at each stage of my career from a Ph.D. student to an assistant professor has decreased. This isn't because I have a reduced work ethic or workload, but because I am far more productive and efficient with my time. The busier you are, the more you will get done; but what will you get done? As a graduate student, your main focus should be your research and getting your dissertation completed so that you can move on to the rest of your career. Here are a few tips at improving your efficiency.

(1) Remember that you are a researcher first. No matter what you want to be as an academic, at least in the Ph.D. student stage you are there to write a research thesis. You may be paid for teaching and other assistantships, but this is not why you were accepted into the program. You should try to be the best teacher possible but not at the expense of being the best researcher possible.

(2) Teaching/office hours. Office hours should be by appointment only, and only during a designated period. For most classes that you will TA, you will be asked to set aside time for office hours to meet with students who have questions about the lecture or lab that they want to go over. For most courses, few students will actually go to office hours unless an exam is coming up or they have a crush on their TA. Rather than setting aside time for office hours in some place other than where you work (such as a library, taking you away from your research), you should ask the students to let you know ahead of time if they will be coming to office hours by making an appointment. You don't want to waste an hour or more of your time waiting for students to show up. At the same time, if you don't have a designated time for office hours, students may think that you are on call 24/7. Giving students the option of reaching you at anytime in your office is a big mistake because they will come by unexpectedly, often when you are busy doing or thinking of research. These interruptions can really hinder your productivity. You will find yourself answering the same questions over and over again to different students, so try to avoid these one-on-one sessions that break up your day. You should be free to set aside large blocks of time for your research when you are not interrupted by undergraduates. The same goes for phone calls; keep these restricted as much as you can. Don't give students your cell phone number! Give them your office phone number only. E-mail should be the first option for students reaching you outside of office hours. Therefore, have designated office hours, but ask students to tell you in advance if they will be coming.

(3) Reverse time line. Find out the time of day when you are the most productive in terms of reading papers and writing. For one week, keep a timeline of your activities during each day, hour by hour. See when you

are doing your best work in terms of your research and when your brain is useless for such activities. If you find yourself an efficient reader with your morning coffee, then you should set aside that part of the day to read papers in your field or your assignments. If you find you write best after lunch, keep that time to concentrate on doing just that. If you find that 3:00, as they say, is "too late and too early to do anything," then keep this time to do things related to nonresearch activities, like reviewing papers or grading exams.

(4) Try to group meetings together. If you have to schedule multiple com-mittee meetings, seminars, and other activities throughout the week, try to have them all on one day of the week. This will leave you with large blocks of time to do research, which is better than dividing your day with distractions. How much can you really get done between 1:00 and 2:00 when there is a 2:30 seminar?

(5) Teach the lab, not the course. The role of many TAs is to organize and teach the laboratory portion of the class. For many classes, the professor answers most of the students' questions about the lecture part of the course, whereas the duty of teaching the lab falls squarely on the TA. In these cases, the TA is essentially working autonomously from the professor's lecture class and isn't responsible for much of the lecture material. Although they may be better directed toward the professor, students will still inevitably ask you questions about the class. You may be unable to answer some of these questions, but you should be able to refer them to the text or the notes. For more elaborate questions, you can refer them to the professor (who is effectively the manager of the class). You may want to say, "I'll get back to you after talking to the professor" or suggest that the students e-mail their questions directly to the professor. If a student gets annoyed with these responses, which they sometimes do, you can say, "I have to stay on top of things with the lab—the prof will be happy to talk to you more about her lectures." Be honest, but do not blow off the students. It is okay to explain that you don't know everything related to this class, particularly if it is not your area of expertise. Definitely make sure that this strategy is okay with the professor. Often they want all lecture questions directed at them anyway. However, in very large classes (>100 students), part of your duty may require you to explain what the lectures were about. If this is not the case, save time by focusing on your most important duties (i.e., helping the students learn your portion of the course).

(6) Get old notes, learn from senior TAs. Preparing to TA a new course, particularly one with a lab, can be challenging. You should begin by asking those who taught the class most recently for their notes, assignments, and exams. (You should not reuse the exams, because chances are the students already have a copy.) This will save you enormous amounts of time in planning out and organizing the labs. If there are senior TAs teaching with you, you should rely heavily on them and their notes and handouts.

Sometimes you can interject new material, but if it saves you time, use the previous versions. Not everything has to be perfectly equal among the TAs in terms of preparing and scheduling. If you have a senior TA working with you and they have prepared notes and handouts from previous years, much of the heavy lifting for the course is already done (see Box 3.1 below for an example). In return for this incredible time saver, you should do whatever other jobs the senior TA wants you to do.

Box 3.1   An example of how to make a teaching assistantship easier

One semester in graduate school, I was having difficulty finding a new teaching assignment when I discovered *Parasitology* was available. Although I did not plan on ever using parasitology in my research, I did enjoy that class as an undergraduate. I also knew that the other TA had taught that class for years and that the professor handled all the lecture material and exams himself. The class size was about 60, and there were four lab sections that I split with the other TA. I asked if I could teach the later two sections which were taught at the back end of every week, and if I could sit in on the first section of the week taught by the other TA on Mondays. On the Monday before the lab, I would help the other TA put out all the specimens and microscopes that the students would be using. Then I would watch as the other TA made beautiful, elaborate illustrations of parasite morphology for the students on the board. I would have helped with this, too, but I lacked the artistic ability and he seemed content to do this. Those illustrations stayed on the board the entire week so that I could use them in my own lab sections. I stayed for the entire introduction for the Monday lab and participated in part of the lab as if I were a student, making notes on his introduction and the questions that came up. Once I got what I needed and felt like I had a handle on the lab, I would leave. This saved the other TA from having to teach me everything separately, and it saved me from having a separate meeting time with him. I was able to pretty easily regurgitate what I had learned to the students without much prep time.

In the lectures, the students learned hundreds of different life cycles of parasites and lots of interesting behaviors and strategies. I learned them, too: I just didn't have to memorize it like they did. If students asked me about a particular parasite, I would look it up in my notes or the textbook and go over it with them. If they had a question, I couldn't answer, I'd tell them I didn't know and that I'd ask the other TA or the professor. They usually didn't mind this short delay. On the rare occasion they did mind, I explained to them that this wasn't my direct field so I wasn't as informed as the other teachers in the class. Undergraduates are more forgiving about this than you would expect. They are not so forgiving if you are rude or dishonest with them.

(7) Don't begin labs with a lecture. First-time TAs often make the mistake of over-researching a subject and regurgitating it to the students as a long, overblown lecture at the beginning of a lab. This takes enormous amounts of prep time and is unnecessary; the students don't want another lecture on top of their course, and the purpose of the lab is to give them hands-on experience that can't be demonstrated in a lecture. Have them form groups and work on projects together, rather than just listen to you lecture. Start with a short introduction of the activities; you can break up the prepared "lecture" into multiple portions of the lab for introducing different sections. If you need to do a lot of background research to understand the subject you are teaching, it's better to turn it into a handout instead of a lecture. This way the students will still be responsible for the material, but you won't force yourself into giving a presentation (your students will also appreciate not having to hear another lecture during lab time).

You can also take the strategy of participating in the lab as one of the students instead of just being the all-knowing all-mighty director of activities. In the you-as-a-student approach, you are introducing a problem that you and the students will solve together. Participate with them by acting more like the keen student who helps everyone else get through the lab exercise, without giving away any answers, of course. In this way, you are part of the team trying to solve the problems that are presented to the group. This strategy will keep the students active in solving the problems, instead of simply relying on you to tell them the answer. At the end of the exercise, you can go back to your role as teacher and summarize what you all learned and how.

(8) Exams. Depending on how much leeway you have in creating exams, you can make grading as easy for yourself as possible by using multiple-choice questions rather than long essays. If you do short-answer questions, an easy way to make grading simple is to make the question worth the number of key points you are looking for. For instance, if the question is worth five points, you can grade the short answer by looking quickly for five key phrases that you taught them and expect to see in their answer. If you only see three of those phrases, then they only get three of the five points. You can also save time by having a few big exams rather than lots of small quizzes or vice versa, depending on your preference and the size of the class. If you prefer lots of little quizzes, be prepared to have students haggling over every grade each time. This might be an okay strategy if you can handle it and you know what you are getting into. Students sometimes appreciate many small exams rather than a few large exams because it keeps them studying throughout the semester instead of cramming a few times.

Try to come up with exam questions throughout the semester as you teach, rather than trying to come up with questions all at once. This will not only save you time, but generally the questions are better if you've thought them up with the material you are presenting rather than having to go back through your notes trying to come up with 60 questions from scratch.

(9) Keep notes. The second time you teach a course is much easier than the first time, but it can be even easier if you've kept good notes. Good note keeping is always a good idea anyway because you'll likely need them somewhere down the line. Try to teach classes you have experience with during semesters that you will be the busiest with your own research. If you know your qualifying exams are coming up in the next semester, you should probably teach a class you've taught before rather than a new course during a semester in which you will already be stressed.

Picking a Dissertation Topic and Writing Your Thesis

Picking a dissertation topic will be up to you and your committee, but ultimately you will (or should) have the control. Pick something that you think can hold your interest for 3–6 years. Choosing an obscure topic that few people outside of your field have heard of will make it harder to sell yourself (getting your papers cited or your talks well attended, etc.) to people in different fields who may be influential in your career later. (Working on continental drift of Africa is easier to sell as "interesting" to most people than, say, the socioeconomic influences of Afghan taxi drivers in Minnesota.) You should also try thinking of multiple levels of analysis in your thesis. Within the work, you should have some smaller-scale detailed analyses that require a level of experience with the minutia of your field that few others will have (e.g., a species description or a mathematical proof), something at some intermediate level of analysis (e.g., genus-level phylogeny or your own lemma based on your earlier mathematical proof), and an overarching synthetic analysis that is both a summary and a culmination of the details of the other chapters (e.g., a biogeographic study or a test of some physical law based on your theorem).

As discussed earlier, a review paper with a hundred or so references makes an excellent introductory chapter for a thesis. This will set a benchmark for the topics of subsequent chapters. Think of your dissertation in terms of publishable units; each of your chapters can be an independent publication. The best dissertations are made up of coherent units that can stand on their own as treatises on particular subjects. Some people are able to publish on a scattered range of topics and simply "staple" these noncohesive units together and present this as a thesis. If your committee will let you get away with this, more power to you; but if your thesis has a title with three colons in it, it is probably not a treatise on a single subject. Having a singular theme, even if it is broad, will help you build a reputation in a scientific discipline and have a solid foundation for your academic research program.

Few people will actually read your thesis in its final form outside of the members of your committee. Your dissertation is not a peer-reviewed publication; it is only considered "published" when the results appear in peer-reviewed journals. Few people, if anyone, will ever cite your dissertation because it is viewed as an "unpublished" manuscript. Even if you can make your dissertation freely available online, you shouldn't do so until you publish the results. Putting it online before it is published will make it too easy for your competitors to scoop your ideas.

Your dissertation abstract will appear on searchable databases so make sure that you spend a lot of time and thought composing this summary. This is the last thing you will write after your chapters are all written up and approved. Breaking down five years of research into 350 words (the typical length requested) will not be easy. Make sure the major hypotheses, tests, and conclusions are front and center. Once this is out there, people will start citing it, so make sure you get on publishing these results in real indexed journals. Citations of your dissertation won't count in calculating your H-index (see Figure 7.2 in Chapter 7).

The students who have the most trouble writing their dissertations are those who try to write it up all at once at the end of their graduate career. It is much easier to write your thesis as you go, submitting chapters as publications along the way. This strategy ensures that you have publications with your name on them by the time you are looking for a postdoc or job. Writing up your dissertation all in one piece at the end of your graduate career means that you will be writing about projects you did several years earlier, and writing should always be done when the work is fresh; it is very hard to keep track of the methods and the data once a project is completed, let alone years after the data was collected. As you are doing a project, write up part of the introduction and methods while you await the results. Relearning methods each time you do a project is an enormous waste of time that can be saved by writing up a detailed method section for each project that you do. You can cut and paste these sections each time you write up a new project, adding more info as you go along.

Choosing a Committee

Your graduate evaluations committee should be as small as possible. You will be asked to pick from two to four advisors aside from your primary advisor to serve on this graduate committee. The members of this committee are directly in charge of evaluating your progress and ultimate worthiness for earning your degree. If the minimum number of people that you need to form a committee is three, then that is what you should have. Additional people will only have additional work and requirements for you: that is not a good thing. Think of them as being your parole board. You will find scheduling meetings between faculty serving on your committee among the hardest things you will have to do for these meetings, so try to keep as many of them as possible within your department, as scheduling meetings that include external members can be quite challenging.

Having a small committee does not mean that you should not or cannot seek out the advice of other faculty members outside of your committee. You should certainly take advantage of the other faculty members at your university; many will be willing to serve roles similar to your committee members (mentoring, reviewing papers, etc.) without actually being on your committee. Faculty don't get very much credit for being on someone's committee unless they are the primary advisor, so you are probably doing them a favor by absolving them of having to attend committee meetings. At the same time, it might be worth adding extra committee members that can help you get funding or lab space. Research space and funds are hard to come by and usually are only provided to students by the members of their evaluation committee.

Ask other students how their committees have interacted with each other. If some potential committee members argue a lot about science or departmental politics, chances are they will spend your already pressure-filled committee meetings bickering. You should choose a group that knows each other well and that may have already served on committees together. It is often a good strategy to pick an identical committee to a more senior and successful student in your lab, although you should be prepared for comparisons to be made between you and them.

Grants and Thesis Proposals

Your time will not only be spent teaching and writing up your thesis but also applying for internal (departmental and university-wide grants) and external grants. Before writing applications for grants, you should write your thesis proposal, and before you write your thesis proposal, your introductory review paper should be nearly complete. One will lead the other, and each is easier to write once the others are complete. From the references in your review paper, you will have a grasp on your field of study that few others have. You will be able to understand the gaps in the field and where the questions that remain unanswered lie. These are questions that you can answer in your dissertation. From this basis, you can write a thesis proposal that you can submit to your committee for approval and that you can modify as grant proposals.

At a minimum, a thesis proposal must include an outline of ideas for each potential chapter of the thesis and a timeline for completion of each project. You should have an abstract for the goals of each chapter and perhaps potential journals where the results can be submitted, as well as a list of meetings where talks based on those chapters can be presented (see Appendix 3 for an example thesis proposal). The more details the better, and this proposal should be updated each year until the final product is completed. The thesis proposal of a first-year graduate student will be thin on results but should be big on ambition. I was once told by a professor that today's students are "smart but not ambitious." It is better to think big than to have your thesis committee thinking up new directions for your overly simple thesis. Your graduate committee might think that your proposal is too simple because you don't have the sufficient background to think up a more ambitious project. As a student you should be pushing the available technology to answer questions in new and interesting ways. Every year the data that will be available to you will change, and so your thesis proposal should change with the times, too. I'd be surprised if anyone has ever completed their thesis exactly as outlined by their initial proposal. However, if you submit a very ambitious proposal, you should be able to defend yourself when asked if you can actually complete it. You will be asked if you will have enough grant money, time, and experience to do what you want to do. If you say "yes" and then don't come armed with some products (papers and grants) the next time you meet with your committee, you will get hammered. I still recommend thinking big; the higher you set the bar, the higher you'll learn to reach.

The thesis proposal is your guide to writing grants, just as the introductory review paper was a guide to writing your thesis proposal. Whether the grant is big or small you can use your outlined thesis projects as a guide. As you refine those projects for the grant, you can go back to the thesis proposal and refine it as well. Typically, your committee, or at least your advisor, will want to see some sort of progress report each year (or more frequently) showing them (1) publications (thesis or nonthesis), (2) grants that you applied for, (3) an updated CV, and (4) your updated thesis proposal (see Appendix 3).

Typically the grants that you apply for in your first year as a precandidate (the stage before you've passed qualifying exams) are based more on your undergraduate grades and standardized test scores. This includes the NSF-predoctoral award. After you have reached candidacy, you will apply for grants that focus more on your thesis proposal and graduate school CV, such as the NSF Doctoral Dissertation Improvement Grant. For these grants, you want to keep an updated CV (see Appendix 2). Ask senior graduate students that have had success to let you see a copy of their grant applications and CV. Granting agencies may ask to see a variety of items in your applications, including everything from publications to synergistic activities

Meetings/Societies

Annual meetings of scientific societies for your field are the best chances you will have to meet and impress not only your peers but also the people who will be deciding if you will be a worthy postdoc or job applicant in the future. More people will likely watch your presentation or view your poster at a national meeting than read any of your publications. You may have the funds to go to only one of these a year, but going to two or three is not a bad idea as they can lead to more options for postgraduate work. Earlier in your graduate career, one meeting a year will be plenty; it is when you are on the postdoc or job market that you want to start putting multiple meetings in your annual schedule. As you become more senior, you will have first dibs on grants that will fund you to go to meetings, but it might even be worth taking on some out-of-pocket costs if the meetings really might pan out as a venue for finding your next gig. Even if you have limited funds, try at least to give a presentation at the most important meeting in your field every year. You will see the same faces and get to know more and more people every time you go. You need to attend annually in order to build a reputation. Maintaining a presence is important if you want to move up to the next academic level in that discipline.

If you go to meetings, ask your advisor or other senior people that you know to introduce you to important people in the field. Make a good first impression, as it will last. Often you will have to introduce yourself again to the same person at another meeting but at least you can say, "Hi I'm X, Y's student, we met briefly at the Tampa meetings." Too many students see meetings just as a time to hang out with friends and not as a time to showcase themselves. Of course meetings are fun and you should hang out with friends, but you should also try to move up the professional ladder, not just the social one. The shy student should remember that there is no name in your field that is too big to approach. None of them were born with Ph.D.s; they paid their dues, and they were all once where you are now. Going up to someone and saying, "Hi, my name is X, my advisor is Y. I really enjoyed your work on [some subject], I work on [some similar subject]. I'm giving a poster [on this day], I hope you can come see it." This exchange sometimes can earn you a beer with someone you admire and perhaps later down the road a collaboration or even a position in his lab. It can sometimes be hard to muster up the courage to approach someone you admire, so if you are too shy to do it yourself, ask someone to introduce you. If you want to meet people who have similar interests and can help you down the line, ask your advisor for a hand.

After the meetings are over, it is wise to shoot a quick e-mail to some of the people you met and had good conversations with, such as: "It was nice chatting with you at the XXXX meetings. As I mentioned I really admire your work and hope that we can collaborate down the line. I've attached some .pdf files of my work that you might be interested in." A short note like this can put you on someone's radar that you may otherwise be left off of. Remember, they've met dozens of new people: what will you do to stand out? Hopefully it won't be your karaoke singing or your drunken rant at the banquet.

Knowing When to Finish

Students waiting to have the perfectly completed dissertation usually take way too long to finish their degrees. There will always be more you can do; there is always another experiment. Never think of your dissertation as being a completed study; there will always be loose ends. Your dissertation is just the summary of the earliest research stage of your career. Many researchers build their careers based on the work they started during their dissertation, and many researchers reach their peak with their dissertation; obviously it is better to be the former.

So when is it time to finish? A Ph.D. program typically takes between 4 and 6 years. You should start making inquiries about jobs and postdocs as soon as you become a candidate. At meetings, you should look at job postings and pay attention to who is advertising postdoctoral positions. Seek them out and ask them what they are looking for in a candidate for their position. You may feel like you don't have enough seasoning yet, but people are often looking for someone on the rise. If you are someone they see as a good fit in their lab, they may be willing to take a chance on you. Your dissertation committee will be a lot more open to letting a few unfinished experiments slide if you have a postdoc or job waiting for you after you finish. On the other hand, students completing their dissertations without a position pending may find themselves with a committee much less lenient about incomplete projects. Essentially, you should not graduate until you know where you are going next, or unless you are sure you can complete your dissertation and land something rather quickly.

The worst thing you can do is stick around the program long enough to watch your dissertation get outdated—and that doesn't take long. Technology takes off so quickly in any field, and new publications move the science along so fast that your early chapters, if done at the beginning of your graduate career, may be nearly outdated by the time you finish. If those chapters are published, they will be of little consequence to your committee, and they probably won't even bother reading them. If you are essentially done with most of your dissertation writing but don't have a job lined-up, you can hang back an extra year while you publish your thesis and look for the next opportunity. Doing this for too long, however, is a bad idea. Very rarely will your committee tell you it is time to go; typically, it is some external administrative source and being on their radar as an extra-term "super senior student" is not a good situation to be in. Most places have some sort of cap on how long students can stay graduate students in good standing. Usually that's something like seven years, and after that you'll have to petition them to let you stay. They might even cut your tuition-waiver so that you have to actually pay to be in grad school instead of getting paid. Don't let it get to this point!

You should attend as many dissertation defense presentations as possible. You will notice that the best ones are fresh studies from students finishing up on time; the worst ones are often from those that stuck around too long. Their talks are dull because the work has grown stale sitting on a shelf, and these super-senior students have a much harder time finding work because of it. Keep the time between when you learned a cutting edge technique to when you present results as short as possible so that your final product is as fresh as possible.

You should also enjoy your "end time" as a student. If you have prepared yourself for the end by writing as you go rather than cramming it all in the end, you should have a little time to enjoy your status as a senior graduate student. Teach the newbies coming in about how to survive in the department. Have some lunches with professors who have helped you along the way and ask them about how their academic careers unfolded. Talk to other students about to finish up since you may find them to be your competition for jobs one day. Leave behind a positive legacy at your institution because you might even return one day as a professor. This doesn't always work out, but as your peers graduate and move on, your name may come up during job searches, and you don't want negative opinions about you to hinder your professional progress. It's a small world, and academic circles are even smaller.

As you finish, be sure to wrap up as many loose ends of your dissertation research as possible; no one wants to a hire a postdoc who will be spending all his time publishing his doctoral dissertation. Postdocs are hired to work on their advisor's projects. Try submitting as many thesis chapters as you can before you start your postdoc. If you don't, you will have to find your own time to work on it outside of your postdoctoral duties. Do your best to publish your thesis within the first couple of years of graduating; with each passing month, your thesis will become less and less relevant. The quicker you turn out the products of your thesis the better off you'll be.

Summary

- As a graduate student you are learning how to become an independent researcher and scientist; that is your main duty and obligation regardless of your future career choice.
- Your first year will be difficult as you get settled in. The best way to learn is from more experienced students and postdocs who are a few steps ahead of you. Start collaborating with them to get some experience as soon as you can.
- Don't say no to any collaborative projects. Some will pan out and others won't, but you don't want to miss the boat on a publication. If you do say yes to a collaboration, you are saying you will work hard on your part, so make sure you gain a reputation as a fast and hard-working collaborator, not as one to avoid.
- Seminars are important for you to attend and learn from. Try to ask questions at each talk, which will help you stay tuned-in the entire time and will build your reputation, preferably as a thinker and not as a pest.

- When giving seminars, remember what you learned about giving a good seminar from those you've seen and try to emulate those talks and speakers.
- Try to learn as many of the techniques and software programs that you will need to complete your thesis early on in your graduate career so that you can start implementing those tools and get the results as soon as possible.
- Ask your advisors how you are progressing; it is the best way to know where you stand. Try to meet with your advisor once a week to discuss your development. Advisors don't worry as much about students who are proactive in keeping them updated on their progress.
- Nothing matters more than getting publications and grants. Try to get them early and frequently, and the rest will come easy.
- If your advisor can't provide any research support, don't just sulk and curse your lot—go out and get your own funds. You can do a lot with a little money, and there should be lots of small grants that you can apply for. There might be scientists out there with similar interests who might be interested in your potential results and who also might be willing to help pay for part of your research. The point is to be proactive to get what you need.
- Try to make the products of the qualifying exams required for advancement to candidacy things that will be useful to you (e.g., written exam = first thesis chapter).
- Be a good citizen of the department: it will pay off eventually. Be good to your peers and volunteer to do things. Being a recluse and not helping out will not endear you to anyone and may cost you valuable professional opportunities.
- Don't sign up as a student for classes that you can audit or TA instead. Only take classes that you really need for your research. Most of what you will need to learn for your thesis will come as hands-on experience. Volunteer to work with people doing interesting research so that you can learn the tools you need to apply it to your own work.
- A great introductory chapter of your thesis is a review paper about the topic you intend to study. Writing this paper helps you see where the gaps are in your field as you gather the literature. Review papers are often very highly cited, so this first chapter can become an outstanding publication.
- Get into a yearly routine of signing up for conferences, working up projects to talk about at those conferences, and then writing up the results into publications after the conference. Pushing yourself into this routine will keep you on your toes rather than resting on your (nonexistent) laurels.
- Teaching is secondary to your thesis research. You might love teaching, and you are getting paid to do it, but don't let it take away more than 20 hours a week of your time. Do a good job as a teacher but make

sure you block out large chunks of time to focus on your primary objective—getting your degree.

- Manage your time wisely. Figure out what the most productive and least productive periods are in your day, and try to keep those times in mind when blocking out time to write or read manuscripts, or to deal with teaching duties.
- When teaching a class, keep good notes throughout the year to help you write exam questions and to help you in case you teach the class again. The second time you teach should take half the time as the first, if you keep good notes.
- Pick the smallest thesis committee possible. This is your parole board: why would you want more people to evaluate you than you need?
- Try to write up your thesis chapters as publications (in terms of length, structure, and formatting) so that you can submit them more efficiently and effectively.

4 Finding a Postdoctoral Position

The Hard Truth About Finding a Postdoc

As mentioned in Chapter 3, the search for a postdoc should begin fairly early. As soon as you pass qualifying exams and you are a candidate, you should start thinking about the next step. Make inquiries in academic circles when you can, and let people know that you will be ready for a position in a couple of years. Of course, not everyone needs to do a postdoc. If you are happy to go on to a nonacademic career or teach at a small liberal arts college where research is not emphasized, then you should be looking for those jobs directly. However, most serious researchers have to do one or two postdocs. Most successful researchers find a job within 1–5 years after graduating, spending that time in one or two postdoc positions. This is prime time for candidates on the job market looking for an assistant professorship (a tenure-track faculty position). Someone who hasn't landed a job in 5–7 years after getting their degree can seem unhireable to a research I university. Just like an old dissertation, people come across as stale; they've played the game and have been unsuccessful in their prime.

Typically, there will be at least two major cycles of postdoc positions opening up every year, although sporadic announcements can happen any time. These two primary periods correspond to the start of the fall or winter semester. These cycles also correspond to the time recent faculty hires are looking for help (the beginning of the fall semester at most schools) or after people find out about their grants (many of them, again, around the

A Guide to Academia: Getting into and Surviving Grad School, Postdocs, and a Research Job,
First Edition. Prosanta Chakrabarty.
© 2012 John Wiley & Sons, Inc. Published 2012 by John Wiley & Sons, Inc.

beginning of the fall semester or at the beginning of the new year). Only the best candidates move on to the prime positions. Just as with natural selection, the postdoc market is all about survival of the fittest, but it is also about survival of the attentive. You can't get a position you didn't hear about.

When starting the postdoc search, the best way to hear about openings that fit you is through word of mouth. The best place to hear about these positions is at scientific society meetings, so be sure to be tuned-in to the conversation among your peers and superiors. Getting yourself known is important in these situations because interested advisors may contact you, although it is somewhat rare to be recruited rather than have to seek out a postdoctoral position on your own. Word-of-mouth job searches often start with your Ph.D. advisor giving a good word to someone they know, so be sure to let your advisor know who you are thinking about contacting.

Someone looking to hire a postdoc is taking a large gamble. Postdocs are expensive (with a salary and health insurance, they can cost upwards of $100K a year; the postdoc only sees a fraction of this as salary), and their brief stay (1–3 years) means that a PI wants to hire someone who can be highly productive right away. Most PIs are mainly looking for a postdoc to get some research project going, yielding publications and grants for their lab.

Start your postdoc search by scanning the various scientific job search websites, such as "Nature jobs" (http://www.nature.com/naturejobs/science/welcome), "Science jobs" (http://scjobs.sciencemag.org), or the Chronicle of Higher Education (http://chronicle.com/). For these searches, posting a job opening can be expensive, so many postdoc vacancies are not posted there because of the expense to the PI. More valuable are those specific listserves and e-mail groups for your particular subfield that your advisors will be able to tell you about.

It is also good to be proactive and contact people you'd like to work with directly, almost in the same manner that you did when you were looking for a Ph.D. advisor. The difference is that there are even fewer people with funds for postdocs than there are with funds and space for Ph.D. students. Everyone would love to have a postdoc, but not everyone has the funding for it. You can scan lists of recently awarded grants (NSF for instance: https://www.fastlane.nsf.gov/a6/A6Start.htm). If a PI received more than $300K in funding, chances are they likely asked for some postdoc funds and may have an opening.

Even if you think it may be somewhat of a long shot, it is worth making inquiries about postdoctoral positions with people you would be interested in working with. Ask them if they plan on having a position available around the time you plan on graduating. If the inquiry is to someone you know well, the e-mail should be short, polite, and straight-forward. If the inquiry is to someone you know of, but who doesn't know you well, it should include your CV, research interests, and link to your website.

Sometimes people stay behind to do a postdoc at their graduate alma mater, and there are obvious advantages to doing this: (1) you know the place, and (2) you don't have to look for housing. There are also disadvantages: staying at your graduate alma mater sometimes makes your CV look weaker than someone who has proven themselves at different locations. (This only works to a point, as someone who has moved every two years for the last ten years can seem undisciplined.) Also, the job of being a postdoc is very different from your previous role as a student. This role shift is harder to adjust to when you are at the same institution or even in the same lab. As a postdoc, you are no longer a student learning at your own pace; you are now an employee who will likely be the workhorse on someone else's projects. That transition is sometimes easier to handle at a new place than the same place where you did your graduate work.

Types of Postdocs

There are several different kinds of postdocs, and you should be aware of the differences and how to apply for them. The most prevalent kind of positions are those in which you work and get paid on a PI's grant. Also common are postdocs on a PI's start-up or institutional funds. Less common are "self-funded" postdocs, where the postdoc brings in his own grant money to work with a PI, essentially of his choice. There are also postdoc positions that are only part-time research; these typically require lab manager duties or teaching classes. Each of these is very different.

If you are working as a postdoc on someone's grant, they will likely want you to fulfill the narrow interests related to that grant. There will certainly be some leeway related to external projects but not as much as other types of postdocs. You are hired to be the workhorse that fulfills and delivers the goods for that specific grant. This means publishing lots of papers, but it could also mean training undergraduates and participating in fieldwork along with the other things the PI promised the granting agency he would do. You will be under the same strain as the PI to complete the work, but your interests may be less invested than those of the PI. The reconciliation of those forces work out only when the deliverables are made in a timely fashion, because if the PI is getting papers and other deliverables, so are you. Essentially, you look good if you make the PI look good.

If you are a postdoc on someone's start-up money, then you have more freedom than the previous type—if the advisor gives it to you. On start-up or other institutional funds, money is typically available to hire a postdoc without restrictions from the department as to what that postdoc does. Often a new faculty member will be burdened with teaching and committees and may hire a postdoc with free range to do projects the PI is interested in but is too busy to gather data for. Typically these projects conform to the

PIs research program, but they may allow the postdoc to be the lead on the direction of the projects. With this type of postdoc, you are more of a research partner rather than an employee. This, of course, depends on the PI. The PI may have specific plans that restrict the postdoc's work nearly as much as if he were working on a grant. These types of postdocs are not always with new professors with start-up funds; sometimes a more established professor will get some institutional money that he can use for hiring. Just as is the case with new professors, the flexibility of your role is based on the PI's discretion, although the more established professor might be under less pressure as he has already earned tenure.

If you are a postdoc who has obtained a training grant, you can typically take that position and money almost anywhere you like. You just need to find a lab willing to take you in and train you. (Normally, you choose this lab before you formally apply for funds.) Depending on the nature of the grant, this kind of postdoc generally gives you the greatest control of how the funds are used. However, the PI who provides space for you will also be expecting some cut of research funds along with their new postdoc. They will also likely expect you to train their students in new techniques. The price for them is that they need to train you first. These sorts of postdocs also show future employers that you can fund your own projects.

Aside from these main kinds, there are other types of postdocs, including some that ask you to teach courses, curate a collection, or be a lab technician. These postdocs that ask for additional work other than just research can be a good fit and training for an assistant professor position. However, because you have responsibilities beyond research, you will have less time to publish than pure research postdocs. In this way, this alternative postdoc can put you at a disadvantage relative to other job candidates. However, these positions can also give you more experience in other areas that will be important as you enter the job market. There might be more flexibility in such a position in that instead of having a PI advising you, you might get a small amount of your own lab space. In these cases, you might be able to stay in that position for a longer period of time than a typical postdoc and transition to a full faculty position when one becomes available.

If you apply for a grant to fund your own postdoctoral position while still a Ph.D. student or a postdoctoral fellow (applying for a second postdoctoral position) at your current institution, you should be aware of some general rules. In these cases, because you are not a permanent employee of your institution, you typically are not allowed to be the lead PI on the grant. Instead you will be a co-PI, meaning that you will have less power than the PI who is in charge of the distribution of funds. If you leave the institution, you might not see a penny from that grant, even if you wrote most of it. It is best to hash out an agreement with the PI about these situations as you write the grant, as he may allow you to keep some of the funds if you get a job in the intervening period of the grant. You will also need to confirm this possibility with your current institution, as it will have to allow the transfer of some of its overhead to the new institution. Surprisingly this isn't as hard to do as you might think, and the granting agencies are usually happy to help out a new investigator.

What a Postdoc Does

You will need to produce from the start. This means trying to get some grant money, but mostly it means you must publish, publish, publish. Publications are the only real measure of productivity. You will also be looking and applying for jobs, and if possible, publishing some papers on your own during this time (typically from work that you started before your postdoc), but your main job is to do the research projects you were hired to do. You might also be asked to train students or even give a guest lecture or two in a class. These are good things to add to a CV, but be careful of being taken advantage of. There should be a limit to these nonresearch activities, and you'll know where this is once you've surpassed it. You and your PI should come to an understanding of your duties and role before you agree to take the position.

A Postdoc That's Not a Postdoc

A postdoctoral "fellow" or "associate" is typically someone who has received a fellowship or other kinds of funding to be a postdoc. A postdoctoral "researcher" might be the same but can mean someone with a Ph.D. who is doing research without funding. In the period after you get a Ph.D., the postdoctoral period, you might find yourself with your degree but without a position. If this is the case, you might stay in your old lab helping your advisor with projects, perhaps earning a small salary as a technician. It is perfectly fine to list being a technician on your CV to show that you were in a new position; however, you are technically not a postdoc because you were not hired into a postdoctoral position. If you are hired strictly as a technician or a lab assistant, it is not okay to call yourself a postdoc on your CV. In these cases, seek the advice of whatever PI advisor you have. If you just received your Ph.D. from New York University and you move to New Jersey and someone offers you some desk space in a lab at Princeton, you shouldn't list a postdoc at Princeton in your professional positions. If you volunteer in that lab in exchange for the desk space, you might get away with listing Princeton as your current address in publications and for correspondence; but unless you and the PI agree that you are a postdoc, you are not a postdoc. I mention this because some people vetting your CV for a job might find this kind of exaggerated position as evidence of dishonesty.

Summary

- Start your preliminary search for a postdoc as soon as you become a Ph.D. candidate.
- Being a postdoc is not like being a student; you will be more of an employee than a student.
- There are several kinds of postdoctoral positions (based on where the funding comes from); find out which you want based on what you want to be in your career.
- Some postdoc positions provide more freedom than others, but the biggest factor on how much leeway you have is the PI.

5 Postdoctoral Positions

The Hard Truth About Postdocs

A postdoc is a hired gun working for a boss; this is a very different situation than a student training under an advisor. A postdoc is typically not there to be trained, but to produce, and the only currency is publications, patents, or grants. You have a boss (the PI), and he pays you and writes your letters of recommendation. You need to keep your boss happy, and although that may not be the easiest thing in the world, you have to remember that you both essentially want the same thing—that is, publications with your names on it.

As a postdoc, you won't have as much time as you did as a graduate student to really take your time learning a project from the ground up, blazing your own trail. When you start your postdoc, you should be ready to have a high throughput of publications in a short period of time. You should also be applying to jobs as soon as you can. The window of time to get a job is short, and it starts closing as soon as you begin. You've made it this far in academia, but the postdoc separates the independent people who can continue on to their own career and those who either leave the field or are resigned to work on other people's projects forever.

Before you actually move to your new lab, ask for a new computer if you can get one. A new computer is part of a fresh start, and you probably don't want to be using your personal laptop as your main work computer. Hopefully you can get a nice quiet area to work in that should be a slight improvement from your overcrowded graduate student space. You should

A Guide to Academia: Getting into and Surviving Grad School, Postdocs, and a Research Job, First Edition. Prosanta Chakrabarty.
© 2012 John Wiley & Sons, Inc. Published 2012 by John Wiley & Sons, Inc.

get the nicest space available aside from what the PI and more senior technicians and postdocs have. Try not to step on anyone's toes before you even get started, but you know what kind of writing environment you can work in and what is unacceptable. You might have to grin and bear it for a while as you wait for a graduate student who is finishing up to move out of their office. Of course, you'll have to remember that you are a temporary employee, so you won't be in any space for a lengthy period of time.

Being the New Guy or Gal in the Lab

Postdoc positions are extremely valuable and usually enjoyable times in your career. You will often hear academics say that their time as a postdoc was the most fun they ever had in science. Most postdocs don't teach, don't take classes, don't go to administrative meetings, and essentially have no responsibilities outside of their research. All of the distractions that you had as a graduate student are in the past; you are finally free to use those skills and techniques you learned in graduate school. Those tools were why your new PI hired you, and you should both want to hit the ground running on this new collaboration. Unencumbered by your previous time constraints, you can start working on projects that were already started in your new lab and come up with new projects that you always wanted to try. You can dust off your new lab's stagnant and forgotten projects and set on them with a new pair of eyes and techniques. These low hanging fruit are a great place to start so that you can start pushing those publications out of the door. You might also have some side projects of your own that you never completed that you can work on with the other postdocs or students in the lab. They can infuse your work with new ideas and interests, too; just make sure they relate somehow to what the PI is doing or that you have the PI's permission to work on them. It's one thing to work on your own side projects, it's another to be working on them using the PI's other personnel's time to do so. I would recommend that you start small and get a nice set of projects going early on so that you can show yourself as a productive member of your new lab. Having a productive start on the PI's projects might earn you time to get some of your leftover dissertation chapters published as well. Make sure that, as you write up old lab projects, you are not forgetting people who were involved in the early stages of that project, perhaps people who were involved before you even got started. There may be members of the lab that worked on that project who have moved on to other things (or other institutions) and might feel slighted if they are not included as an author or not acknowledged somehow. This should primarily be the senior author's responsibility, but it is a good idea to cover your bases.

Build collaborations with your lab group, particularly the students and other postdocs. They will be excited to have this new productive person (that's you) available to bounce ideas off of and to help them with their projects. Take advantage of that new guy vibe to make a good impression by working hard and helping your new labmates.

At the same time that you are getting close to your new lab, you might feel isolated from the rest of the institution. You won't be on the same e-mail lists as the faculty and students at first, so you might not hear about a cool seminar or guest speaker. Try to get on some of these e-mail lists, but you will also want to avoid some of the others that have nothing to do with you (e.g., class announcements, building-code violation warnings). Eventually you might find yourself working with people outside of your lab as well, but keep in mind that your priorities lie with your PI's lab.

When you are starting off, try to be as organized as possible. Graduate students can afford to be inefficient but postdocs can't. If you are learning a project in a different field than your past experience, you will probably need to play catch-up on the readings for that area of research, so use a bibliographic index (e.g., EndNote) or some other way to keep track of what you've read. If you had wished you used a data management program or some filing system for your papers in graduate school, now is the time to implement those changes. You have a fresh start, take advantage of it.

Advisors

The postdoc and postdoc advisor relationship is a strange one. You both have goals and deadlines, but they do not always line up, and this can cause friction. It should go without saying, but you should insist as a postdoc that you are on papers that you write or contribute work to. Sometimes there is a discrepancy between what you think your efforts are worth and what your PI sees; typically this doesn't become a problem unless you are not publishing enough. Try to see eye-to-eye with your advisor and keep a good channel of communication open. You should meet with your new PI at least weekly, but hopefully more often. Unlike your graduate school advisor, your postdoctoral advisor is your colleague and collaborator.

Often postdoctoral advisors are happy to drop their name at the end of the author by-line on publications, even if they contributed little more than space and funds, but space and funds are essential. You may be frustrated at a PI's contribution, but you should remember that he is essentially your boss and should have his name on papers coming out of his lab. Usually this isn't an issue, but if the PI is paying for your salary and for the project, he should be acknowledged as an author. The PI may decline authorship but that should be his call. Ideally, all of the authors worked directly on getting the project done or written up. If the PI is going to get new grants, he will need to show that his previous grants were successful by listing publications he contributed to, so the more you help him with publications, the better his chances at the next grant. The first author (i.e., lead author) should be the person who wrote the manuscript. The last author (i.e., senior author) position usually goes to the PI who funded the project and edited and directed the work. It isn't cool for the PI who didn't do the work or the writing to be first author, and you should challenge this if the PI requests it. You need to show these first-authored publications in order to get a job, and to show that you can be a lead on a paper. However, the PI might ask to be a corresponding author, and this can be justified because he is the permanent employee at the institutional address, whereas you are not (see Chapter 7, Figure 7.1). The corresponding author will receive all of the e-mails and questions about the paper, so it is best to assign this role to the person who will permanently be in the lab. You will probably have a new mailing address and e-mail in a couple of years, so you might want to pass on being the corresponding author. However, there are some institutions that count how many papers you have as corresponding author when looking through job candidates, so it might be to your advantage to be the corresponding author if you are also first author. Again, these things are best resolved with a frank conversation.

If you are publishing a paper that is your own work, and that was carried out without the use of resources paid for by the advisor, you should be allowed to publish it as the sole author. These kinds of papers could include comment pieces or editorials but also work that was completed before you joined the PI's lab. It is probably best to work on these at home or during off hours to avoid conflict. Remember, the PI is paying your wage and you should feel obligated to dedicate at least 35–40 hours per week to his projects. Some advisors will even allow you to have entire days out of every week to work on these. You should come to an understanding about ownership of projects. If you came up with a project and carried out part of it in the PI's lab, you should be allowed to continue those projects on your own in your new job, but you should probably recognize the contributions of your former PI. This might mean including him as an author on papers in which most (but not all) of the work was done with your own funds at your own job. By hiring and training you, the PI has created a new competitor, so stay on good terms when you leave. It is usually to your mutual benefit to continue collaborating, rather than working in competition.

As a postdoc you might find yourself working a 9-to-5 schedule. A 35- to 40-hour work week may not sound like much in an academic setting, but when you have no other obligations aside from the research, it can be plenty. Remember that the extra 20–40 hours you put in during graduate school were probably due to teaching and other duties not related to research. A postdoc only has to focus on the research. Every PI is different, but as long as you are completing the work the PI requested, it is unlikely that he will push you to work longer hours. He may request that you work essentially the same hours as he does so that you can be there when he needs you; so if you only worked nights as a grad student, you might have to change your schedule.

Conflicts between a postdoc and an advisor often arise from the postdoc doing things without discussing them with his advisor. There are a million ways this can happen, but it often starts with an innocent misunderstanding. Most of the misunderstandings relate to the postdoc unknowingly overstepping his bounds and the advisor feeling slighted because of this breech of power. I know of a time when a postdoc was helping a student from another PI's lab learn how to work on some DNA analysis. The student would come to the postdoc's lab and use some of the reagents and equipment. This doesn't sound like a big deal, but when the PI found out he was quite upset. "Who is paying for the reagents? Why wasn't I told someone new would be working in here?" This situation obviously could have been avoided if the PI was notified. If the PI says "no," usually the postdoc can show the student in another lab, as long as it doesn't take away a big chunk from the postdoc's time. (Usually PIs don't mind you helping others, but some folks are really touchy about these kinds of things.) Unauthorized purchases or vacations taken without notifying the PI in advance are other common points of friction. Spending the PI's money without his okay is always a big no-no. The PI has to watch his budget very carefully, and if an expense comes up that he didn't authorize, there could be a real issue. If you have a hard time reaching your advisor and you need to order something critical, send him an e-mail explaining the situation. Good communication is the best way to avoid issues like this.

You may be befuddled by your advisor's behavior and reactions to something that you think is relatively minor. (Read the Assistant Professor section to see what all the stress is about.) You should look at these issues from the advisor's perspective. He was once a hotshot researcher like you, a problem-solver, a hands-on whiz kid who did things on his own. Now, he has to rely on others (like you) to complete projects while he writes grants and papers and handles a million administrative duties. That loss of direct control over projects means that, in his mind, he might have done some aspect of a particular project differently, or he may have forgotten how long it takes to trouble shoot a particular problem. If the PI is kept out of the loop, he may feel underappreciated or he may be secretly steaming that some project is not working well. Don't let these issues fester; cut them off by talking to the PI as often as possible about what is going on in the lab. The PI might be feeling out of touch or like he is losing control or respect; the only answer is communication, communication, communication.

Running Out of Time

One of the odd sensations that a new postdoc will feel that you didn't feel as a graduate student is the heavy weight of passing time. As soon as you begin a postdoc, you will be under the pressure of deadlines and publishing, but you will also feel the pressure of having just a few years to find success on the job market or to find another postdoc. (Remember most postdocs are only 2–3 years long.) In the narrow window of your postdoc, you will have successes and failures, and you will feel each sensation more acutely as you get closer to running out of funding. Welcome to the life of surviving on soft money. Some academics build entire careers on soft money, living grant to grant with no assurances that they will be able to support themselves as scientists once the current grants are up. Funds for a postdoc sometimes run out before that postdoc has found a new position. At that point, you will be lucky if someone gives you a desk to sit at; you are essentially a volunteer, and at worst a loafer. Some former postdocs take jobs as secretaries or get paid hourly as research assistants just so that they can keep a university address (and health insurance) while they apply for new jobs. Sometimes the new job doesn't start for a few months, and the postdoc has to scrounge for menial work to fill in the gap. The headache of changing health insurance and working essentially in the same position you had when you first got your bachelor's degree can be soul crushing. If you get into this situation, try to keep your spirits up and your productivity, too. The last thing you want is a gap in your publication record, so work doubly hard to get some research hours in. This might mean some late nights and weekend work outside of your regular work hours. Do what it takes to get out of the hole; otherwise you will join the ranks of those many lost souls stuck in those positions permanently. The only thing to do is stay positive, use your new temporary position to look at different kinds of research that can be done in your field, look at it as a new opportunity to learn new techniques, make new collaborators, and hone old skills. Leaving science for a period of time is a hard thing to do, but sometimes circumstances beyond your control force you out into the periphery. It is very difficult to return to active research once you have left a position; the longer you are away, the harder the return. The loss of postdocs who leave the field is one of the greatest sources of attrition among academics, particularly women (see Chapter 8).

The best way to leave a postdoc is by finding a new position, so be sure to apply to things regularly, keeping your ears and eyes tuned-in to news about job openings. The search for a job usually kicks into high gear in the second half of your postdoc when the pressure is really on. Hopefully you have a good cushion of papers by this time so that you can devote more time to this search. At some time in the second half of your postdoc, you should give a seminar that is essentially a practice job talk. You can talk about your dissertation research, but you should also spend a great deal of time talking about your recent work. Your PhD thesis and postdoc might be disparate topics, but try to have a central theme. Giving a talk early in the second half of your postdoc to your department will give you an opportunity to introduce yourself to some of the folks you may not have met or interacted with much to that point in the larger department outside of your PI's lab. This might lead to additional collaborations, but you should also get advice about the talk and what to change if you do turn it into a job seminar.

Summary

- A postdoc to postdoc advisor relationship is more like an employee/ employer relationship but is hopefully also one between collaborator and colleague.
- The relationship between postdoc and advisor sometimes might feel like one of "mutual underappreciation," but the best way to avoid hard feelings is to keep the communication channels open and use them often.
- Your PI will be happy if are a helpful member of the lab and a leader. This includes publishing tons of papers with his name on it. Everything else can be forgiven if you are a real publishing workhorse.
- Start off your postdoc the way you wished you had your first year of graduate school in terms of being organized and efficient.
- Publish the easy stuff first. Work on some of the projects that never got completed in the lab or side projects that you need some help from your new labmates. (The easy stuff doesn't include thesis chapters that aren't part of the PI's projects.)
- You should always be aware of how much time you have left in your postdoctoral contract and try to maximize your productivity by being very efficient during your stay. The postdoctoral period will be the only time that you can focus completely on research without the distractions of administrative duties and teaching. Take advantage of it!

6 Applying for Jobs

Applying for Jobs

It is never too early to apply for jobs; the worst that can happen is that you don't get an interview. You should keep an up-to-date CV that you fine-tune for each application and an introductory letter that is modified with each package that you submit. You should apply to jobs you don't even really want, just for the practice. When the right job does come up, you will have more confidence that your application has gone through the ringer a couple of times. (As with all writing, something that you put away and revise and look at with fresh eyes usually gets better with each revision.) You may even get some "practice" interviews at jobs you might not be all that interested in. The more interviews you have, the better you will get at answering questions and performing as a candidate. At the same time, you must find a balance. Your postdoc will not go very well if you spend the entire time applying for jobs; your advisor won't be happy, and your productivity will suffer. The first job application you write will take a long time, but each subsequent one will essentially be a revision of the first, so they won't always take as much time to write. A postdoctoral advisor should be understanding and allow you time to work on job applications and prep time for interviews. Sometimes you only have a couple of weeks to prepare a presentation for a job interview, and if you

A Guide to Academia: Getting into and Surviving Grad School, Postdocs, and a Research Job,
First Edition. Prosanta Chakrabarty.

don't have something ready to go, it can be rather stressful getting a nice job talk together.

Even as you are applying to jobs, you should be thinking about how you are building an independent research program; start thinking like a PI and how to make yourself hirable. If you have a fair number of publications (10+, but there is no magic number, and quality matters too) and perhaps a grant, you will likely be a strong candidate for jobs. I often see people who would likely be strong candidates for a job who don't even apply because they talk themselves out of it. These pessimists say it would be a waste of time because they think they are poor candidates or because they don't exactly fit the job description. Job descriptions are often vague and written or modified by an administrator; you should apply to jobs that you would be interested in regardless of the exact wording of the announcement. A strong candidate who doesn't fit the initial vision of the hiring search committee may win out against a weak field. In some departments, particularly large ones, a small subcommittee writes a job advertisement that is then approved by a subdepartment, but because the entire department likely votes on the candidates, someone with broader skills other than what was initially intended might ultimately be hired (e.g., an animal behavior person might be hired as a neurophysiologist). A lot is at stake for a university when a job comes up. A failed search may result in the loss of a job line where a position is removed from that department permanently. There are different personalities and desires among different members of a hiring committee; you may strike a chord with one member who then sways the rest of the group. The point is that you really never know, so you might as well throw your hat in the ring. You should always give yourself a chance if you really think the job is one that you would want. There are not that many opportunities to get a job, let alone a job you would like; you don't want to kick yourself for not applying after you learn that someone less qualified was hired. I've seen lots of searches where I thought the short list was rather weak only to hear that some stronger applicants had not applied. The regrets of not taking a chance at applying to a job can last a lifetime—so apply, it can't hurt.

Depending on the job market for your field, you might apply to 10–20 jobs per year, although the variance is highly dependent on the field. In a more specific market (curator of fishes), you might only find 1 or 2 jobs offered each year, whereas a bioinformatician might apply for 70 jobs because his research is so broad that he might fit in a department that advertises for anything from a computational biologist to an engineer.

Applications

When an advertisement for a job you are interested in comes out, you should e-mail the people who are running the search. You might be so bold as to flat out ask what kind of person they are looking for, or you can say that you intend to apply and ask if they can provide more information than is available on the advertisement. Try to find out what they recommend in terms of strategy. Sometimes you can get some interesting tidbits of information. If the advertisement is written very broadly, as they often are, you might find out that they are really looking for someone with a certain skill set. If that is the case, you should emphasize those skills in your application. If you don't know anyone on the search committee, make sure you are polite in your e-mails and make sure you have a link to your website (that should have your CV on it) at the bottom of the e-mail. Assume that everyone at the institution will know of every inquiry you make. You don't want to have a reputation as being pushy or sneaky right off the bat.

The application is typically in several parts. There is the introductory letter (see Appendix 4 for a cover letter), a statement of research interests (Appendix 5), a curriculum vitae (Appendix 6), and a statement of teaching interests (Appendix 7). Some jobs may ask for additional materials, but these four are the typical core items. Some places may be looking to hire someone with some additional skills, and you can add a statement to the package fine tuned for that aspect of the job. (I've included that extra statement for a curator in Appendix 8, and show how you can slip that extra bit into your cover letter in Appendix 4.) You want to keep your application package relatively short but thorough, and you want it to stand out among the other applicants. You don't want to be gimmicky about it, but a little color or even some figures won't hurt and neither does a nice presentation (quality paper, clear well-organized text, etc.). If you use 8-point font in a 20-page, single-spaced document, you are putting yourself at a disadvantage versus someone who submits a more visually compelling application. Of course, content is more important than presentation, but you want to present your application in a way that will highlight your abilities. The search committee may get a hundred applications and may have a limited time to scan each file. Make sure you have clear subheadings so that your major points will catch the reader's eye. Most of the things in your application will be just a few pages, but your CV can be somewhat longer (still no more than 15 pages). If you submit a long CV, again you should have clear subheadings so that if someone is flipping through the pages they can tell where to focus.

The first page of your introductory cover letter should be written on the letterhead of your current institution, and the letter itself shouldn't be more than a couple pages long. In the letter, you should be bold about stating why you would make an excellent candidate and an excellent person to hire. This is not the time to make humble statements: you should exude confidence. Instead of "If I am lucky enough to be interviewed for this position, I look forward to telling you about my research plans," you should write the following: "As a member of the faculty at X-University, my research program will include both ecological and evolutionary components. As you can see from my CV, I have strengths in the areas of research that this position advertises: bioinformatics, quantitative genetics, and microevolution." If you make bold statements like this, you should make sure your CV matches up and is organized in a way to showcase these strengths (see Appendices 5–9). Remember that the search committee will have very little time to go through each application in detail, so make sure you use clear formatting and clear straightforward writing.

At the bottom of your cover letter, you will list three or four references who will write letters of recommendation on your behalf. Make sure these are people who are attentive to their e-mails because a university administrator might be contacting them to get a letter. Sometimes you find out you are in the running for a job because letters are requested from your references. For some jobs, you need to have these letters submitted with your initial application. If you apply for many jobs, this can be a real burden on your referees so also make sure that these are folks who understand your plight as a job seeker. They should also be people who know you personally and not just professionally. These people should be your graduate school advisors and postdoctoral advisors but perhaps also other established mentors that have known you for a while. If you aren't sure that all three referees will be able to submit letters on your behalf in time, list a fourth alternate that you are sure will be able to. Unlike in graduate school admissions, people applying for jobs probably won't have their application tossed out because of a missing reference, but why risk it? Just as in grad school admissions, the committee is searching the letter for negative comments, but they are also looking for positive comments about your congeniality and professionalism. A nice comment like, "Jane is the hardest working person I know, and she is perfect for this position" can make a candidate move up the list of applicants. After asking your advisors for dozens of letters, send them a little something to thank them. These are the people getting you a job: they deserve some credit, even if it's just a thank-you note and some candy.

Job Interviews

Great, you made the short list. Everyone in the field will hear about it. Everyone at the institution you applied to will be looking at your website and Googling you and trying to look up your Facebook profile; such is the vetting process of academia.

Job interviews are pretty standard. Basically, they are weeding out the hireable from the unhireable interviewees and ranking those hireable. I once heard that there are four kinds of candidates: qualified nice people, unqualified nice people, qualified jerks, and unqualified jerks. Try to be in the first category.

After you apply, you might get on the long-short list; this is the trimmed down list of all applicants selected by the search committee as viable candidates. You might get a call from someone on the search committee if you make the long-short list; this quick phone interview might be just to determine whether you are a lunatic or still interested. Usually the search committee then presents this long-short list to the department, who then pick two to five candidates to interview (usually three). Once the list is compiled and the candidates are invited, a schedule is drawn out. Very rarely are two candidates on campus at the same time; usually each candidate is there for two nonoverlapping weekdays. During those two days, you will give an hour-long research presentation and perhaps also a teaching presentation (a talk that you would give as a lecture for a class you might teach), or a "chalk talk." A chalk talk is supposed to be a session where you stand up in front of a group of faculty and present your ideas about your future as they ask you questions for each slide. The chalk talk is essentially a formal group discussion session with props. You might not have to make a formal presentation for this, but I would (see Appendix 9 for how I would make the first few slides). The chalk talk presentation won't need many slides, and you shouldn't have a lot of text, just talking points like you would in a journal club discussion, except now the topic is you. The rest of your time is one-on-one interviews with students, faculty, and administrators as well as some group interviews. The days will be long; after interviews, there is usually a dinner or a social where people can chat you up even more. Interviews can be really fun (you are the center of attention), but they are always long and a little stressful. Be charming, humble, and in teacher mode. Explain everything clearly and in simple broad terms, giving enough detail based on who you are speaking with. Once the interview list is set, it can be anyone's game. Just like in the play-offs for baseball, there might be a favorite going in, but that doesn't mean that team will definitely win the World Series. Act like it is your job to lose. Be nice to everyone you meet, and try to be calm and collected. The more you talk to people, the more comfortable you will get, especially after your talk is done.

After a few weeks, you will hear back either that you got the job or you didn't. The person who is the favorite gets called first. If they end up turning it down, they may offer the position to the next best candidate all the way down the list of acceptable candidates (some applicants will be voted as unacceptable by the faculty after their interview) until they get someone or decide that no one else is worthy of the position.

Job interviews are stressful, but you should also see them as golden opportunities. You will get to meet a wide range of people from administrators to graduate students but mostly the faculty that will be your future peers. Even if you don't get the job, chances are high that you will run into these folks again in this small academic world. Take advantage of your interview time; there are few chances in life where you get to sit with other researchers for 30 minutes or longer and talk about your research and theirs. Some interviews are set up as one-on-one sessions, others are group sessions where you are essentially being interrogated. One-on-one sessions give you much more freedom to take the conversation in a direction where you are most comfortable and mistakes aren't heard by all. (At least not right away.) One-on-one sessions with faculty also give you an opportunity to ask questions about someone's research and interject how you think you can collaborate with them. You can direct questions toward your research interests at first, but as time runs down, it is okay to ask questions like "What is it like to live here?" You are interviewing them to an extent, too. If you get offered the job, they will want you to take it. During these private interviews, you can even mention things that you just learned during other one-on-one sessions. "I hear that you guys are looking to expand the department to include Math." These interviews are all about survival so don't bring up any social conflicts in your current lab or bad mouth anyone you are not getting along with. If the interviewer knows someone you work with, only say positive things, even if the interviewer is goading you on.

Interrogation-type group interviews (like the chalk talk) in which you are ganged up on by several faculty members at once can feel like you are being judged by a Ph.D. committee all over again. (Which is part of why you go through qualifying exams in graduate school: it's practice.) These can be nerve-racking, and no one performs at his best at these times; they will find a gap in your knowledge, and you will make mistakes. The trick is being comfortable but not cocky, being smart about the subjects you know best and humble and inquisitive about subjects you don't know as well. Most questions will likely be about your talk, but some will also be about your publications. Make sure you know your own work inside and out, particularly methods. Keep your cool and keep in mind that they want a colleague who will not only be their equal but who will add something new, fresh, and exciting to the department. If you think you can pull it off, sometimes getting a little joke in at the beginning of the interrogation can lighten the mood. These little quips can really break the ice and put you and the committee at ease. If you are smiling and comfortable, you will come across as calm (you could also come across as an airhead if the situation is one in which you were asked a serious question and you blow it off.) My favorite way that I've heard of a candidate breaking the ice was during the beginning of a group interview when a faculty member asked a difficult question right off the bat and the candidate answered with "Whoa! What is this, a job interview or something?" (That doesn't sound like it would work, but it did the way it was delivered.) You don't want to plan anything in advance, but try to lighten the mood with a joke when you can. A bad joke will only hurt your confidence and darken the mood even further, so only try humor in this situation if you have the goods to pull it off. A witty reply and a sly smile to a tough question are okay if you have a good straight answer coming right after. On one of my interviews, I was given a rather softball first question about a new exhibit hall that was being built at that institution. I was asked, "If the exhibit designer came to you as a new hire to help with the exhibit, what would you do?" I replied, "I would say, I'm sorry but I'm far to busy to be dealing with that." I was joking of course, and I gave a real answer afterward, but that little wisecrack did a lot to clear the stiff air.

If you can find out who the other candidates are, it might be to your advantage to go first, particularly if someone who has similar research interests is also a candidate. I've seen searches where the postdoc and the PI in the same lab were interviewed or two postdocs from the same lab. In these cases, it is to your advantage to go first so the second person looks like the copycat. You don't want to sabotage your labmate, but you do want to distinguish yourself and your research from theirs. Likewise don't say anything negative about any of the other candidates; this will only come back to hurt you.

If you can get your interview schedule in advance, it is to your advantage to learn whom you are meeting with and research their background a bit. This will help you find mutual interests with these people and lead to more fulfilling conversations. Typically, you can get the schedule a week before you actually visit the school. I would look up everyone on the schedule and other people you are likely to encounter (e.g., students, administrators). I would prepare a small pocket notebook with notes about each person (nothing negative in case you lose it) and their work, along with a schedule and facts I wanted to drop every now and then. I would keep this notebook in my suit jacket pocket at all times in case I needed it. A notebook is a lot more discreet and academic looking than using your smartphone; using your phone makes you look like you are distracted. If someone asks you a question you don't know a good answer to, get their name and e-mail them the answer later. Usually these interviews are only two days long, so you will have plenty of missed opportunities.

Depending on how the job search is run, the job candidate's presentation is typically done early on. Be sure to have your computer and presentation as carry on luggage because you won't want to deal with lost luggage when you are supposed to be giving a talk. Keep a copy of your talk and some of your publications and notes on a thumb drive, just to be extra safe. You should wear nice dress clothes that show that you care about your appearance. You can wear a muumuu to work every day once you get the job, but during the interview you don't want people focusing on how poorly you dressed. You will likely meet up with well-dressed administrators, and you shouldn't feel underdressed in their presence. (There is a reason they call them "power suits.") I've seen one male candidate give a talk with a see-through, short-sleeve shirt that looked so ridiculous I couldn't take anything he said seriously. Bring something in your bag in case you spill something on yourself; you don't want to feel self-conscious for your interviews because there is a giant coffee stain on your pants. The interviewer will think you are self-conscious because you are intimidated by the questions not the stain, so bring a little stain remover with you. A water bottle always comes in handy, especially when you get a scratch in your throat during your talk.

On long interview days, you might sometimes need a little break and you will not have the opportunity to take one. You obviously can't cancel a meeting; instead, do what you can to make things more comfortable. If you are in a one-on-one interview and it is nice outside, ask the interviewer if they wouldn't mind chatting outside. Don't look at your watch and ask about the next meeting unless you feel things are really going beyond the allotted time. If you need a break, ask to use the restroom, or ask the interviewer if you can grab some coffee with him. Usually the one-on-one sessions are not longer than 30 minutes, so be diligent about your time, although this is generally the local host's duty.

There is typically at least one social activity, sometimes dinner or a party where people not on the interview list get to meet you. You can have an alcoholic drink if the rest of the group is having one, but you shouldn't feel obligated to or obligated not to; just go with the flow without overdoing it. Try not to be too awkward; if you are a shy person, find a seat somewhere where people can sit around you and you can be most comfortable. If you are outgoing, you should be yourself but don't overdo it. Doing a keg stand might impress the graduate students but likely not the other faculty. Likewise, some folks are turned off by someone who curses, so try to refrain from vulgarities until you are hired. People will engage you and chat you up about your life and work, and even though you are tired and want to sleep, try to talk to everyone as if they were the dean.

Like old people, a job candidate should never pass up a bathroom: you may not see one again for a while. You should ask the person who is creating your schedule to give you at least a half hour to set up for your talk; this will give you time to use the restroom and also plenty of time to get your talk ready. Avoid technical difficulties with the presentation, microphone, and laser pointer; make sure everything is ready to go before people start filing in. Check for conversion issues between different presentation formats if you switched computers. Be sure you know how to operate the overhead lights and how to advance slides. If you can't figure out how to use the computer remote, the audience might wonder how you'll ever master the coffee maker in the graduate lounge.

You usually don't get much choice over where you get to stay, but if you do, try to stay at a hotel as close to the institution as possible. This way if the person in charge of picking you up is late, you can call the head of the search committee and arrangements can be made. In your little interview notebook, keep the phone numbers of important contacts in case you get in trouble. Typically, the head of the search committee is the person entrusted with your well being. You will be exhausted after the first day; once you decompress (i.e., collapse in the bed) try to take it easy. The next day will be another full day of meetings, so don't go clubbing or boozing it up too much at the hotel bar. A better move is to unwind and go over who you met and who you will meet the next day. Try to remember names of the folks you met in case you see them again on day two—they will certainly remember you.

Research Presentations

Begin your talk with something like the following: "Thank you so much for this opportunity. I'm really delighted to be able to speak to you about my research. I'm very happy to have the chance to meet all of you and excited about a chance to work at this institution." Keep the intro short and simple. Show that you are grateful and that you are excited about the opportunity. Don't introduce yourself in terms of where you have been because the person introducing you will cover that. Likewise, even though much of your work has probably been collaborative, you should stress the "I did this" instead of the "we did this" whenever it makes sense to. Try not to emphasize that something was your Ph.D. research because that will make you sound too much like a novice. Instead, direct your talk in terms of research projects in which you were the lead, or did independently, and try to build a nice story arc where there is some large question that you answered using multiple tools. The department wants to hire someone with skills who can do new things and collaborate with others while expanding their own research program. Build your talk around your big picture research question, and make sure this is broad enough for everyone in a wide audience to understand its importance. Make sure you state that big picture research question at the beginning and end of your talk. Try to emphasize your skills and cite your publication year and date at the bottom of the data slides (e.g., Matamoros and Chakrabarty, 2009, Cladistics). If you are the fourth author and you cite Smith *et al.* 2010 (and you are not Smith), it won't look like your work, so mention during the slide, "This is work I've done with Leo Smith at the University of Chicago." You can talk about your collaborative projects, but you want to emphasize your first-authored projects. You also want to get into data slides rather quickly. If you are giving a 40-minute presentation, you should be giving data slides before you hit minute 10. The early introductory slides should state somewhere that "my research program covers this [broad area of research]. I'd like to talk specifically about this [more inclusive area of research] today." Throughout the talk, you should keep this big picture as the main theme while also emphasizing the methods and skills that you may have that the other candidates may not.

Your presentation should be a coherent story, but it should also showcase your diversity and your plans for the future. Have a few slides at the end after your conclusions (but before acknowledgments) to talk about your goals and plans for research directions at this new institution. Show that you plan to take your research to the next level and beyond; think big and show them why they need someone like you.

Keep the presentation at about 40–45 minutes. That leaves time for a five-minute introduction by the local host, and assuming the talk starts ten after the hour, it will allow for you to end before the end of the hour and leave some time for questions. The worst thing to do is go over your allotted time. We have all been to talks that run over the allotted time, and the longer it goes, the more frustrated the audience will become. God could be giving a talk, but if he goes five minutes over the allotted time people will be yawning and slinking toward the exit. Keep your talk within time and leave plenty of time for questions. How you handle questions can make or break a job interview. I've seen candidates who were the clear favorite going in lose their shot by being disparaging or aloof when answering a question. I've seen a candidate become a slam-dunk choice by answering a cruel and overly critical question from a particularly gruff faculty member with a simple "well, we are doing the best we can, and of course we are working to improve on these data." Be humble and be prepared for questions; the best way to do this is to practice your talk as many times as you can in front of your friends and colleagues. (Read the section about giving seminars in Chapter 2 for additional pointers.)

The best practice for a job interview is giving invited talks at other universities and research institutions. If you have friends who recently got jobs, ask them to invite you to give a talk to their lab or department. You will get a similar schedule of interviews with students and faculty after your research talk without the stress of being a job candidate.

End your talk with an acknowledgment slide that includes the selection committee and other people who may have helped you at that institution. You want to be a good fit within the department and a little bit of name-dropping doesn't hurt. End with something like "thank you again for the opportunity to talk to you today. I look forward to meeting with the faculty and students later today. Thank you." Don't end with "and now I'll take questions." This call-for-questions results in holding off applause, whereas "thank you" will encourage it.

After your talk, you will be rushed off to more interviews or lunch (often pizza with students), so keep the rush you feel after the talk going for as long as you can. If the talk doesn't go as well as you had hoped, try to shake it off; all is not lost. If you need to take a bathroom break, take a few minutes to get your thoughts in order. Use the interviews after your talk to preemptively strike against some mistakes. "Geez, I really blew the question by Dr. Fink. I should have said that [insert correct answer here]."

Usually you get an "Oh, don't worry" as a reply, but when Dr. Fink brings up your bad answer at the faculty meeting, you might have someone there to mention that you knew the better answer and realized your mistake.

Once you've left and arrived safely home, send an e-mail message to the people you met thanking them for their hospitality and wishing them good luck on the rest of the search. If you really connected with a potential collaborator, make sure you mention in an e-mail that you really would like to work together in the near future no matter the outcome of the search.

Typical Questions You Are Asked in a Job Interview

You should have summaries of your research or introductions that last 30 seconds, 2 minutes, and 10 minutes. You will be introduced to people in the hallway on your way between interviews. People who are only told that you are a candidate for a position may ask, "so, what do you do?" You should answer, "I'm an XXXXX. Pleased to meet you. I work on the physics of swimming at XXX. I hope I get a chance to meet with you later." Other times you will be at a lunch with graduate students stuffing their faces with pizza (the only sustenance they've likely had in days) and you should lead off by introducing yourself again and talking about your research interests and then asking them to do the same. As they mention what they do, try to interject encouragement or potential collaborative future projects. The graduate students are typically not a very important voting block, but they can sway the decisions of those who do have a vote.

You will be repeating your research goals over and over, so make sure you have them clear in your head. You will also be asked often about (1) your first research grants that you intend to apply for, (2) where you want to take your research program, and (3) what you see yourself doing in five years. Know the answers to these and base them on the needs of the department and the job that was advertised (being honest of course). If you know the department needs someone to teach genetics and lead a lab, say, "I plan on teaching genetics and applying to NSF for funding to get a molecular lab started here. I see a lot of opportunities to collaborate with the faculty here to add a molecular component to the department."

You might be asked, "Do you have any questions?" Ask harmless questions like, "Can I see the lab space" or "Will I be given funds for renovations?" Avoid questions about salary and start-up; these are negotiated once you are made an offer. You should be agreeable to every condition as a job candidate. If they show you your office and it looks like a janitor's closet, just smile and nod as if it were the Taj Mahal. Only after you are offered a job are you in a position to negotiate. If you are asked by an administrator whether you have questions, ask questions that lead to showing off your interests: "Are there funds to purchase materials for teaching? I'd like to purchase a thermocycler to show undergraduate students how to do some hands on genetic work."

Here is a list of typical questions you might get. Each question is really an opportunity for you to showcase your research and experience.

(1)   What are your research goals?
(2)   Do you have an idea of what courses you would like to teach here?
(3)   What skill sets and backgrounds are you looking for when hiring students and postdocs? How will you recruit them?
(4)   What is your best work?
(5)   What are the most recent *Nature* or *Science* [or top journals in your field] articles that you've found interesting?
(6)   If you did not work on the particular subject that you are trained in, what other field would you work in?
(7)   What is the big picture of your research? How do you explain what you do to your grandmother?
(8)   What other non-dissertation, non-postdoc projects have you done? [i.e., what are your side projects?]
(9)   What kind of funds will you apply for? [Know of several specific grant opportunities.]
(10) Where do you see yourself in 2, 5, 15 years?
(11) Do you see some collaborative opportunities with some of the people you've met here? [Yes, you do!] What kind of researchers do you normally collaborate with?
(12) What kind of lab space/lab equipment will you need? [Don't say anything too crazy unless you say you are planning on getting the money yourself to provide it, or they already have it. You can negotiate once you have the job to get those things.]
(13) What society meetings do you normally attend?

Not Getting the Interview/Job

There is no better pat on the back then scoring a premium job interview; there are no greater disappointments than not getting an interview you thought you deserved. There is nothing worse than spending weeks working on an application and perfecting it only to hear that you did not get an interview. It's okay to scream and get upset and take some time off to cool your heels. But after the shortlist is declared, there is really nothing you can do about it but sulk. Even if it was your dream job and you thought it was perfect, all is not over. You will get other interviews and you will end up at the job that is right for you. No matter how bad the job market is, people always seem to end up at the right place eventually. Some people take more time than others to get to where they should be.

Remember that everyone has to pay their dues at some point. Some pay more than others.

Getting an Offer

You got a job! The search committee chair has called to congratulate you. They will send you an e-mail or letter soon with more details about the offer. They may fly you in for a second visit to talk about your needs and check out your lab and office space again and to negotiate start-up (see the next chapter for more about these negotiations). The ball is now in your court. If it is the job you want, awesome; the decision will be easy. If you are not sure that's okay, they are the nervous ones now. If you are fairly confident that you will get another offer and that this isn't the job for you, let them know once you've made up your mind and talked to your advisors. You don't want to string them along, but you also don't want to make a hasty decision. If you string them along for too long, they may miss an opportunity to hire their second choice and that can breed resentment. You also don't want to have a reputation for turning down jobs. A lot was invested in this search, but ultimately you have to make the right decision for yourself.

You might find yourself with two simultaneous offers (always a great situation to be in). If you can manage it, you can get each institution to play off of each other's offers to get more than you were originally offered in salary, space, and start-up. Don't be a jerk about it, but get what you need to ensure that you will be successful and be in the best situation to get tenure. Remember that these people will be your colleagues and peers. You don't want to burn any bridges before you've even started working with them.

You may also find yourself waiting on a call from your preferred institution while you are being hounded to make a decision from your second choice that has already made you an offer. Be honest to both parties. Let your second choice know that you would like to hear back from the other institution (without mentioning that you would prefer to go there). You can tell your first-choice school that they need to let you know as soon as possible if you are still in the running. Let them also know the situation and that you will be forced to say yes to the other institution by some given date. Make a date if possible, even if the other institution hasn't given you a deadline. If your second choice gives you an ultimatum that you have to take the job or leave it by a certain date and you still don't know the outcome of the other search, you might have to say yes to number two. You just can't let a bird in the hand go. There might still be some wiggle room if number one comes calling later and the best thing to do is be totally honest with both parties. If you want to be at your top choice, then find a way to say yes to them while still keeping your integrity.

Summary

- Apply to jobs even if they are not exactly the perfect job description; they might end up being the perfect job once you go there on an interview. Also, the more you send out your application, the more revisions it will go through; so, by the time your real favorite comes up, your application will be ripe and ready.
- Be smart about your application. If you really want a job, stay on top of things. Find out if all the materials were received and mail extra copies to members of the search committee.
- Typically, an application consists of a short but direct cover letter, a statement of research interests, a focused CV, and a teaching state-ment/philosophy. Make sure these are written clearly and with sub-headings and titles that emphasize your strong points.
- If the job description includes an extra element like curation or teaching, then you can add an additional statement to your application to show that you mean to take that part of the job seriously as well. That extra effort may land you the job.
- Prepare as much as possible for the job interview, not just the job talk. You should research the people you are meeting on your schedule, especially if you could potentially collaborate with them.
- Enjoy the interviews. It will feel like you are teaching, except the subject is you. People will be asking you about your research and your interests. You are the star of the show: eat it up.
- Take advantage of the opportunity to pick the minds of these potential future colleagues. You will rarely have an opportunity to meet with re-searchers one-on-one to talk about science like this. They will appreciate your enthusiasm.
- Make sure your talk hits on big picture points. This might be the broad-est audience you've ever presented to. Can you convince a dean and a new graduate student that your research is interesting and important? If you can, everyone in the middle will think so, too.
- Talk about your research as if it was all recently completed. Don't mention that something was done for your Ph.D. thesis; it will make it sound like a thesis defense more than a job talk.
- Include slides about your future plans and goals in your new depart-ment. Mention the kinds of grants you would apply for and the kind of collaborative lab you would run. Show them why they need you.
- Have talking points that you can use if you meet someone briefly or if you have lengthier conversations. Ask them questions about what they do, so that it isn't only about you. Treat everyone you meet (stu-dents, non-academics) as you would the most powerful person you would meet.

7 Assistant Professor

The Hard Truth of Being an Assistant Professor

Yes! You've been hired! Nothing feels quite as good as that sweet period of time between that first congratulatory phone call and the time you make your first visit as a new employee to your very own job; yes, your very own professional position—congratulations, Professor. Ahhh, you rock. Enjoy that sweet delusion while it lasts. Soon you'll begin teaching, writing grants, and furiously writing papers with the hope that it will be enough to be granted tenure. You might find yourself saying, "This is not what I signed up for," except it is. I hope this chapter helps the reader expect the unexpected. I also hope it gives the forward-thinking graduate student insights into what your advisors went through and what you will be going through one day.

A Guide to Academia: Getting into and Surviving Grad School, Postdocs, and a Research Job,
First Edition. Prosanta Chakrabarty.
© 2012 John Wiley & Sons, Inc. Published 2012 by John Wiley & Sons, Inc.

Negotiating Start-Up

Let the wooing begin. Now that you've been hired, the folks at your new institution will want you to take the job. You've finally got the upper hand, take advantage of it: you are free agent LeBron James (or at least Chris Bosh). Typically for a job at a research I university, the hiring institution will offer to fly you and your significant other to visit what they hope will be your new home. The important part of this visit is negotiating start-up packages and finding a place to live. In both cases, you should talk to the most recent hires at the equivalent level that you were hired (i.e., other assistant professors).

You will spend more time speaking with the administrative faculty (Deans, Chairs, etc.) on this visit then you will at any other time in your career (unless you become an administrator yourself). Before these meetings, ask the three or four last hires to send you their start-up wish list (see Appendix 10 for an example). If the last three hires received $400,000, ask for $500,000. You probably won't get that big of a bump up in funds, but if you ask for the same amount as the last person, do you expect them to give you more?

You should get more than the last guy, but every situation is different. If you are hired as an assistant professor sharing a joint molecular lab, you won't have to spend the $150,000 on new lab equipment that someone starting a lab from scratch would have to, and the administrators know this. In the case of a shared lab, it would be wise of you to ask what the equipment needs are of the lab and to tack those items to your own start-up list (see the shared items in Appendix 10). If you are hired as a philosopher, or a theoretician, you likely won't need, and you won't get, nearly the amount of money that an experimental physicist who requires expensive equipment will receive. Your start-up funds should be equivalent to a medium-sized federally funded grant in your field. Your start-up funds aren't a welcoming gift either: it's a loan, a loan from the mafia. If you want to stick around beyond the pre-tenure stage, you will be expected to repay that sum and then some. The only way to repay it is through the overhead on a grant (so get crackin' on writing one). The relief of getting your first big grant is indescribable.

If you have a second visit (the start-up negotiation/looking for housing visit), you will have a chance to state your case for start-up, but you will likely not get to negotiate much. (Unless you are coming from an equivalent assistant professor position.) Typically, you will talk with a chairperson, but it is a dean or vice-chancellor who controls the shots. You might meet with these folks, too, but these meetings are often perfunctory "chat with the new guy/gal" meetings. You may even want to bring your spouse or significant other to these meetings, although you may feel more comfortable without them there for the negotiations.

Typically, there isn't much wiggle room in negotiation of salary. Normally, you get a little bit more than the last hire, but this varies widely across institutions (particularly private vs. public institutions). Your prior experience matters, but I'm assuming you are coming from a postdoctoral position and not another assistant professorship. Unless you have a competing offer (another job offer), you are pretty much stuck taking what they offer. In some cases, people with more teaching experience get a higher starting salary, so keep all those official records of your teaching hours from graduate school. In negotiating salary, you should not take less than a recent hire in an equivalent position with similar experience. Fight for what you think you deserve. You might be so happy to have a job that you let salary differences slide, but eventually this discrepancy in pay will get to you and you won't be able to do much about it later. If you have a good argument, you will likely get what you deserve, particularly if the person you are negotiating with can't justify why someone else should be getting paid more than you for an equivalent position. (This will require you to ask your new peers what their salary is. Most will be forthcoming, if you tell them why you need to know.) This is also a good opportunity to negotiate a yearly raise. At most major institutions, a 2–4% yearly raise is promised but not guaranteed. If there are rough times with budget cuts, these raises are the first to go. If you are savvy enough, you might ask how these raises work at your institution and perhaps negotiate a sweetheart deal that guarantees you a bump every year. (Good luck with that.)

Be careful not to be too aggressive in negotiating; you don't want to burn any bridges you haven't crossed yet, particularly over salary. You haven't been hired yet, and although they want you, it isn't unheard of for an offer to be pulled even at this late stage. Remember, you haven't signed any contracts yet. Reneging a job offer at the last minute is rare, but when it happens, word gets around and reputations on both sides can be tarnished.

Negotiating salary is mostly about getting what is fair and maybe a little about pride; negotiating start-up is a different story. Try your best to figure out what it will cost to get the equipment and personnel you need to complete the projects that will get you tenure. The people you are negotiating with want you to get tenure; explain to them that you are asking for these things to put yourself in the best position to get tenure. Explain that you also intend to get a grant, but that for the first two years while you apply for grants you will need certain things that will keep your lab rolling, getting preliminary data for those grant applications.

You may also be asked if you want to receive a 9-month appointment or a 12-month appointment. The difference between these is simple: do you want to be paid 9 months out of the year or 12? Of course 12, right? Wrong, at least in this day in age. The reason for the 9-month appointment is that it allows you to request summary salary on a grant. Summary salary is funding you request on a grant to cover your salary (for 1–3 months) so that you can work on your grant full time in lieu of working for the university. Of course you are still actually a university employee, but this gap in salary allows you some extra funding, a little incentive to get a grant and get essentially a bonus in your salary. So you should always opt for the 9-month salary. Remember you will get paid the same amount over the year. If your salary is to be $63,000 a year you will still get $63,000 a year in a 9-month appointment versus a 12-month appointment. The benefit is that if you get a grant, you can ask for 3 months of additional salary (at $7K a month, that's an extra $21K a year). Many institutions also incorporate this summer salary into your start-up for the first 2 years. If you take the 12-month option, that means you cannot ask for summer salary and you are greatly reducing your potential pay. In addition—this is the strangest bit—if you are on a 9-month appointment, you can still ask for that money to be distributed over 12 months instead of just 9, so you won't have to worry about not getting a salary over the summer.

Some of the things that you can negotiate won't have anything to do with money. If you can get it, ask for a teaching abstention your first semester. If you need to set up a lab and put in a grant in your first semester, it will be a lot easier if you don't have to teach at the same time. Asking for a semester off from teaching the semester before you go up for tenure is a good idea, too. You might not get it, but this is a way to get something that is just as valuable as additional funds.

If you are hired to work in September and your current job ends in July, you may want to ask for the department to cover the cost of COBRA (Consolidated Omnibus Budget Reconciliation Act). COBRA is a government-subsidized program that helps to cover your health care costs during periods between jobs when you and your family are not covered. This $1000 or so is typically covered by your new department. The other cost you will want covered that has nothing to do with research is moving costs. Moving can cost you thousands of dollars, and depending on the institution, may be something that you pay up front and get reimbursed for. You may also have to get multiple bids for this expense. Many universities will pay for a professional moving service to pack up your old place and move it to the new one without you having to do anything. Sometimes they will even move your office from your old university. Typically, there is some cut off on how much they can give, but it doesn't hurt to ask if you need more. (See the bottom of the list in Appendix 10.)

If faculty at your institution are asked to provide funds for graduate student stipends, you should ask for these funds in advance and perhaps ask for funds to be permanently allocated for a student's stipend so that you will never be stuck without a teaching assistant for a class or a graduate research assistant. You will be surprised what you will be able to get, so it is always worth asking. When negotiating, you can also ask for other faculty members to make a case on your behalf. The director of your research unit could fight on your behalf to justify the need for a research assistant, technician, or postdoc to a dean or chairperson if you don't feel confident that you can make the case yourself. It would benefit the head of that unit to have that new position as well, and it is easier to negotiate a new position if you are saying it will improve the unit as a whole.

Remember when coming up with your start-up wish list that you should be able to justify everything on there. You want to avoid the awkward situation where the Dean asks you why you need an espresso maker in your lab. If you are asking for lab supplies, consider that most institutions often have discounts with major retailers, so the administrators may reduce the amount they give you based on these discounts. It is unlikely that anyone will ever check whether you purchased the exact things on the list; the total amounts and justifications are more important.

Once all the negotiations are done, you will get an official offer letter that spells out your salary, space assignment, start-up funds, teaching abstentions, etc. This is your contract, so read it over very carefully and make sure everything makes sense to you and fits with your expectations. If there is something you don't understand, ask the Chair or Dean to explain it. If something was written incorrectly, ask them to send you a corrected version. This letter should also tell you about the department's expectations for you for tenure and other reviews. The expectations of the university for you to get tenure will likely be illustrated in a vague way, something like:

> "As a faculty member of the X-University, your work responsibilities will be 50% research and 50% teaching. We expect you to have an active research program that brings in major national funding and involves graduate students and produces high-quality publications. You will be expected to teach one 3-hour course a semester and a graduate seminar at least every other year. Excellent teaching is expected at X-University at both the graduate and undergraduate levels."

You will have to figure out the finer details about the department's expectations on your own, but these broad strokes should be something you strive to achieve.

Again, pay close attention to this offer letter; make sure you have multiple copies. You may need it when you ask for that teaching abstention the chair of your department didn't know you negotiated. Administrators also change; the person you negotiated your start-up with may be gone, but this contract with the university still stands.

Other Considerations Before You Actually Start Your New Job

Find out what is expected of you for tenure. If you need at least 15 publications, two graduate students, and at least one large federally funded grant, then get cracking. One of the things I highly recommend is that, as you are

TAXONOMIC REVIEW OF THE PONYFISHES

Prosanta Chakrabarty[1,2] & John S. Sparks[1,*]

[1]American Museum of Natural History, 79th Street at Central Park West, New York, New York 11004, USA.
[2] Current Address: Museum of Natural Science, Louisiana State University, 119 Foster Hall, Baton Rouge, Louisiana 70803, USA.
* Corresponding Author

Figure 7.1 Typical author by-line, showing current (new job) address and previous address.

finishing up your postdoc or previous position, start putting your new address on papers. You should also have the address of the institution where you did all the work, but put a second institution on your papers with a "current address" as your new job (see above). Many institutions will count all your papers where that institution appears in the by-line, so it is to your advantage to have as many papers as possible with the new address. If you are not starting your new job for two years, it is a little disingenuous to have your new employer as your "current address," but if you are starting in a few months and the paper won't come out until you are at your new desk, then you should most certainly have your "current address" be the address of your new institution (Figure 7.1).

The period before you move is also a great time to make sure you have some good collaborative projects with the folks from your old institution. It will take a while for you to get settled into your new job, so gather as much raw data as you can at your old job so that you can write up the results at your new job. (Don't forget the people at your old job who paid for these data and helped you when listing authors!) You will have a lot of time early on at your new job to write, so having data to publish will ensure there isn't a long gap in your publication record.

The most important metric that people will use to measure your publication record is your H-index, and you should start thinking strategically about getting this index as high as possible. The H-index quantifies your publications and the citations of those publications. To measure your H-index, see the tutorial below (Figure 7.2). Only indexed journals (i.e., journals that are catalogued and searchable in Web of Science, Scopus, or other bibliographic databases) are considered, so book chapters and publications in regional journals are often not factored into your score. (Google Scholar is great for looking up papers, and it does calculate citations for book chapters and more obscure references; however, it is typically considered too broad and inaccurate to calculate your official score.)

The H-index is much maligned for essentially boiling down careers into raw numbers. Two people with H-indexes of 15 might have careers of very different value and success, and your peer academics will know to

Calculating Your H-Index

H-index = 5

1   **(15 Citations)** Chakrabarty, P., Davis, M.P., Smith, W.Berquist, R., Gledhill, K., Sparks, J., and Frank, L. (2011) Evolution of the light organ system in ponyfishes (Teleostei: Leiognathidae) *Journal of Morphology* 272, 704-721.

2   **(10 citations)** Harrison, I.J., Chakrabarty, P., Freyhof, J., Craig, J.F. (2011) Correct nomenclature and recommendations for preserving and cataloguing voucher material and genetic sequences. *Journal of Fish Biology* 78, 1283-1290.

3   **(7 citations)** Schaefer, S.A, Chakrabarty, P., Geneva, A., Sabaj-Perez, M.H. (2011) Nucleotide sequence data confirm monophyly and local endemism of variable morphospecies of Andean astroblepid catfishes (Siluriformes: Astroblepidae). *Zoological Journal of the Linnean Society* 162, 90-102.

4   **(6 citations)** Davis, M.P., Chakrabarty, P. (2011) Video Observation of a Tripodfish (Aulopiformes: *Bathypterois*) from Bathypelagic Depths in the Campos Basin of Brazil. *Marine Biology Research* 7, 297-303.

5   **(5 citations)** Chakrabarty, P., (2010) Genetypes: a concept to help integrate molecular systematics and traditional taxonomy. *Zootaxa* 2632: 67-68.

(5 citations) Chakrabarty, P. (2010) The transitioning state of systematic ichthyology. *Copeia* 2010 (3):513-514

(2 citations) Ho, H.C. Chakrabarty, P., Sparks, J.S., (2010) Review of the *Halieutichthys aculeatus* species complex (Lophiiformes: Ogcocephalidae), with descriptions of two new species. *Journal of Fish Biology* 77:841-869.

(2 citations) Chakrabarty, P., (2010) Status and phylogeny of Milyeringidae, with the description of a new blind cave fish from Australia, *Milyeringa brooksi*, n. sp. *Zootaxa* 2557:19-28.

(2 citations) Chakrabarty, P., Ho, H.C., Sparks, J.S. (2010) Review of the ponyfishes (Perciformes: Leiognathidae) of Taiwan. *Marine Biodiversity* 140:107-121.

(2 citations) Chakrabarty, P., Chu, J., Luthfun, N., and Sparks, J.S. (2010) Geometric morphometrics uncovers an undescribed ponyfish (Teleostei: Leiognathidae: *Equulites*) with a note on the taxonomic status of *Equula berbis* Valenciennes. *Zootaxa* 2427:15-24.

Figure 7.2 Your H-index is measured by the number of citations you have for each of your publications. The inventor Jorge E. Hirsch, defines it by this equation, "A scientist has index h if h of [his/her] N_p papers have at least h citations each, and the other ($N_p - $ h) papers have at most h citations each." You can use the Web of Science to calculate this for you or you can do it on your own. Just list your publications from the highest number of citations to the lowest as in the above examples. The highest number of citations you have with that equivalent number of publications is equal to your score, The H-index above is 5 because I have five publications with at least five citations. Someone with five publications each with five citations would also have an H-index of 5. Someone with 1000 publications all cited just one time each would have an H-index of 1. The fastest way for my H-index to go up to 6 in the above example is if both of my publications with five citations is cited one additional time each.

look beyond this number. However, if your H-index is below 5 and you have been a publishing academic for 15 years, it is hard to say that you've been publishing important papers. That score means that, in 15 years, you have only 5 publications that have been cited 5 or more times. You might have 30 other papers, but they were also cited 5 times or less. On the flip side, how do you deny that someone whose H-index is 50 didn't have an influential career? Although still a rough metric, the H-index is one that administrators and your peers will find a valuable measure in your evaluations. Look up the H-index of some of your peers to get an idea of how you stand.

The most important thing you can do to improve your H-index is to publish in indexed journals and journals that have high impact factors (i.e., a value also calculated in bibliographic databases like the Web of Science, that is based on the average number of citations for articles in that journal). Although the impact factor of the journals themselves is not integrated in your H-index score, the chances you will be cited is greatest in journals that have higher impact factors. A somewhat sneaky way of getting your score up is self-citation. But, if you are not citing your own work, why would anyone else? Don't rely on this too much; it is relatively easy to calculate someone's score without self-citations.

I would also recommend that you have your office ready to go as much as possible before you arrive at your new job. If your office needs to be painted, get that done before you arrive so that it doesn't distract you while you are getting settled in. Also make sure you have a computer with internet, a phone, and a desk all set up and ready to go before you arrive so that you can hit the ground running.

How to Spend Your Money and What to Expect

Although you will make a request for start-up funds that will be pretty exact (an Excel spread sheet with several hundred items is not uncommon; see Appendix 10), you don't have to spend the money exactly as you requested. Start-up funds at a public university often cover the cost of overhead for hiring someone. That means if you hire a postdoctoral fellow for $35,000, you won't have to worry about the extra $75,000 in overhead that goes into health insurance costs and other costs associated with that hiring. The most important thing you can get with your start-up is probably a postdoctoral fellow. A postdoc will do the work that needs to get done as you go up for tenure without being hindered by committees or teaching assignments that you will have to deal with. If your funds are limited, a technician, someone skilled in the lab but without a Ph.D., might be a better option for you (see below).

One of the unexpected costs that can really add up is renovations. If you are moving into an older building and need a sink, a lab bench, an extra electrical outlet, or a light fixture, you might be shocked at the estimated cost. You would be wise to find out how much facilities services at your institution charges for these things before negotiating start-up. A $30,000 sink is not unheard of (happened to me). If it is already past the start-up negotiation phase and an unexpected renovation cost comes up (like, say, a $30K sink), you might still be able to request departmental funds for this work.

Your office will be your new home for the next few years (maybe the next 40 years); you will basically live there (you might even die of old age there). You really only get the discretionary funds to put in a carpet or nice lights when you get start-up money. Ask for your office to be painted and whatever else you want before you arrive if possible. If you want carpeting or new lights, now is the time to ask because you will rarely see money again with the flexibility of start-up funds. (Try using NSF money to put carpet in your office and you might find yourself looking for a new job.) Although start-up finds are flexible, be very careful to follow the guidelines of your institution; you don't have carte blanche to spend as you please. However, if you are going to be spending 40–90 hours a week for the next few decades in that office, you should think about what will make it a nice environment for you to work in comfortably.

You can also save a lot of money by purchasing items or getting them free from other labs. Instead of buying a brand new desk, chairs, and shelves, head to the surplus department that many large institutions have and see what is available. In some places, administrators change computers every year, so you might get a free or cheap perfectly functional year-old computer to give to a graduate student or to run analyses. Funds for new computers, renovations, and service packages for equipment are sometimes difficult to justify on grants, but they are perfectly fine start-up purchases. Take advantage of the flexibility of these funds, particularly if there are no overhead costs charged to them. Remember that if you want to hire someone on a grant, you basically have to ask for twice as much as their salary to cover health care and other indirect expenses. If you don't have to pay overhead on your start-up, then a technician making $20K will only cost you $20K.

One of the best money savers is asking for price quotes before making purchases. Whether it be lab equipment or renovations and whether it costs $5 or $5000, you should get a quote before you buy. Only suckers pay the sticker price displayed on major suppliers' websites; with a quote, and the backing of your institution's name, you can save between 10% and 50% of the sticker price if not more. Almost every major supplier of equipment has a sales rep that takes orders and applies discounts for most institutions. Find out who this is from your colleagues or set up your own contact. Even if you are at a smaller institution, you can make contact with one of these reps simply by asking for a quote before making your next purchase. The person replying with the quote should apply a discount, and this can be your contact for future purchases. Asking for a quote can save you many thousands of dollars in the long run.

The First Year

There are two major paths that you can take in your first year: (1) hire no one and spend your time publishing papers that have been sitting on the back burner for ages, or (2) hire lots of people to generate new data and results. I personally prefer the first option, but somewhere in between might be the best route for most new hires. Just remember that every new person you hire will need your help and possibly your hands-on training. If you do hire people, make it top heavy; hire postdocs before technicians and technicians before students so that the greener ones can learn from the more experienced ones without too much of your direct intervention necessary. If you are the only senior person in the lab, then you will have to be more hands on with a student than you may have anticipated.

Start-up funds usually have an expiration date; you are often required to spend all of it in the first two years. Remember that these are funds to get you on your feet until you get your own grant. Sometimes the expiration date is stated pretty strictly in your contract, but informally you can get away with spending it for another year or two. Find out from other recent hires how long they were able to go beyond the expiration date, but be careful because technically the department can rescind the funds if you go past this date. If you have some flexibility with this date, I would suggest that option 1 above is best because you will have plenty of time to hire postdocs and to bring in graduate students after your inaugural year. Let this year be your time to publish those single-authored papers that you didn't have time for before. That side project that you worked on for your dissertation but never wrote up: write it up. I would spend as much time writing as possible and not only papers but grants, too (see below). If you have a nice series of first-authored papers come out early in your new job, people will start to take notice. Remember that these papers will likely not come out for at least a year as reviews, and rewrites will take a while. I would shoot for the low-hanging fruit first; try to get as many publications as you can in the bank as possible by publishing in journals with quick turn-around times. You don't want to be stuck trying to catch up on pubs in the 11th hour of your tenure clock. Getting a series of papers out as early as possible will give you some breathing room in your second and third year to publish the bigger, higher impact papers once you have your lab in full swing. If you are in a large department, it may be difficult for your colleagues to distinguish between your papers and their relative scholarship (except by the impact factor of the journals). However, your colleagues will be able to judge the difference between two publications and 15. Don't be shy about writing comment pieces and review articles either; as long as they are peer-reviewed, they count. Comment pieces show that you have an important perspective on your field and that you can publish those views. There is a limit to this, of course; if you have 20 articles and they are all comment articles that include no original science, you will most certainly be judged as having a far less competitive package when compared with someone publishing high-impact papers based on original work. You will come across as an armchair reactionary, whereas the other person is doing "real" science.

Some departments also look to see how many of your papers are published with your previous postdoctoral advisors or colleagues. They may consider these of lesser value or may not count them at all in your tenure package. Make sure that the ground rules are clear to you well before you go up for tenure review. In Appendix 11, I have included the kind of summary presentation of your tenure package that one of your colleagues might present to the department when you go up for review. These no-frills presentations show just the facts about your career, so be sure you know what categories are most important for these reviews well in advance.

Hiring People (Technicians, Students)

If your start-up package allows you to hire people without paying overhead charges, make sure you save enough money to do so. Saving money can be difficult if you are at an institution that expects that you to spend all of your start-up within the first year or two. If this is the case, you have to spend money quickly and you should hire someone quickly. If you don't have the funding for a postdoc, hire a technician. A technician can be someone without a Ph.D. but who has enough training to gather the data you need without much supervision. A technician is much moreso your employee than a postdoctoral fellow; a postdoc is more of a colleague. A postdoc is someone you expect not only to gather data but also to write papers with you. A technician will be paid less but should be generating data for you at your request. In hiring a technician, I would look for someone who is already trained and who is available at your home institution or near-by institution. This will keep you from having to deal with them moving and having to figure out how to get settled. You can also hire a technician for a short period of time, perhaps even just a few months. You will have a hard time finding a postdoc willing to join you for a period of less than a year. Also, a technician is a better option than a postdoc if you don't have your lab set up yet. A postdoctoral fellow is looking to be in that position for 1–3 years, and they will have a hard time padding their CV if all they are doing is ordering supplies and arranging equipment. Although this role can be perfectly fine for a postdoc, it isn't the best use of their time or your money.

Finding the right people to hire depends on your personality type. I was always told never to hire someone you never met and never hire someone for a guaranteed time longer than one year. You want the option to get rid of someone who isn't working out after a year. You can do this by being clear that there are certain things expected of the person you want to bring in (i.e., three publications, one conference talk). Do your homework in advance; ask the colleagues of your potential hire about what it is like to work with him/her.

Conferences are a great place to find someone to hire. As you know, academia is a tough place to find a job, but at least it is a relatively small world. Sticking up a flier recruiting students, postdocs, or professionals usually attracts a great deal of attention; if not, you might be at the wrong conference. If you are looking for a postdoc, ask your colleagues if they have a student finishing up his Ph.D. that they are keen on; ask other graduate students, too. I've found a great way to find good recruits is to sign up to be a judge of student talks at conferences. Here you can see master's student's talks that might turn out to be good Ph.D. candidates, and you will likely see late-stage Ph.D. students who might be good postdocs.

If you find a good person, recruit them hard. Don't be shy to let them know you want them to join your lab: they should be flattered. Good people are hard to find. For a postdoctoral candidate, invite them to give a talk at your institution (another nice thing about start-up is the flexibility to pay for things like this). Find out from them when they plan to finish their current postdoc or Ph.D. If they are really good, then they will be worth the wait if you can't get them right away. Unfortunately, if you are hiring someone to work off of start-up funds, you might not be able to hold on to the money before it expires. At some institutions, you can put all the funds you expect to use toward a multiyear postdoc or technician's salary in a "lock box" (i.e., separate account) so that you technically "spent" all the start-up money within the allocated time but you really still have funds to hire someone later.

I also like to find out how much of my potential postdoc's dissertation is published. Make it clear that you expect them to be paid to be doing collaborative work with you not just publishing their Ph.D. thesis. It can also benefit you to let them take a bit longer to finish up their Ph.D. before starting their postdoc if they are submitting papers from their thesis during that time. That way they can concentrate on new projects once they make the move to your lab. You can also negotiate to give them time to publish their own work while working with you, as long as you come to a mutual understanding about what you expect from them.

There are also logistical issues you will have to deal with when you want to hire someone. Depending on where your funds come from, you might have to submit an advertisement for the job even if you already know who you want. Make sure you deal with all these hurdles first before you promise anyone a job. There are also numerous listservs and websites for every discipline that postdocs and faculty use when looking to hire someone. If you find someone from one of these lists, make sure you talk to them and get to know them as much as you can before you actually make them an offer.

Finding Graduate Students

A colleague once told me, "A bad graduate student might not only ruin your career, they can ruin your life." The rule of thumb here may be to choose carefully and to do your research. Graduate students need your attention and supervision; if they don't get what they need from you, they may rebel. If you are starting a new job, the admissions committee should look favorably on admitting your first-choice recruits. Some tenure committees may expect that you graduated a student by the time you go up for tenure. (If you go up for tenure in your 5th year, that would mean you should find a graduate student in your first year, which means you should recruit before you even start your new job.)

I know some brilliant scientists who are awful advisors. I've known advisors to yell at their students, going so far as to call them "idiots" or "stupid." This is totally unacceptable. As an advisor, you need to nurture a student and help him out of trouble; there is never justification for name calling and degrading. Students will make mistakes—sometimes huge mistakes—but as long as it was not done with malicious intent, they should be forgiven.

I'm not saying students need to be coddled; you can still be tough with them, but you need to remember to treat them with decency: students are people too, after all.

Sometimes it is better to stay out of the way of a graduate student who has an independent spirit; other times it is best to intervene if a student is floundering. The most important thing is keeping an open line of communication. An informal weekly meeting over lunch or coffee may be sufficient. Some students may need more help, and just as the case was when you were a graduate student, most of the help may not need to come from you. Provide other outlets for advice to your student, perhaps postdocs or senior graduate students in your lab or in another lab. Ultimately, they are responsible for their own success but it is your responsibility to do your best to help them find that path to success.

Graduate students are not free labor; they are trainees and future colleagues. You can have them clean out tanks and feed mice but not to the point where they are unpaid technicians. If they are actually getting paid as a part-time tech, don't work them beyond the amount of hours they are getting paid for. Usually the opposite is the problem. Some students think that if they are getting paid as a technician or research assistant for their advisor, then they can slack off hours to work on research. Make it clear what your expectations are, and make them fill out hourly time sheets if that is what works. Don't let resentment build; if they are not putting in the hours for their paid job, it might be because they are focusing on their research a little bit more that week and maybe that's okay. There needs to be a happy balance where both of you come to an understanding of how things are going to be.

Each student is different, and what worked with one may not work with another. A student will eventually have to convince you that he is an independent thinker and problem solver (and problem finder) before he can graduate. In the early stages it may seem like the student is not making the progress you expect; it is best to give him time to find his footing. Do your best to try to remember your early-stage graduate career and all your false starts and mistakes. Let students figure out some of their own mistakes and let them learn some lessons on their own without letting them get too deep in a rut. Eventually it will be up to you to get them back on their feet (see example below).

While you may be focusing your attention toward your own projects as a new professor, your fledgling graduate students are just learning how to become scientists. If you write up their data without their input, you are not helping them learn how to communicate as scientists. You might be advancing your own career, but you are not helping them on their own paths. The most important thing you can teach a graduate student is how to communicate with other scientists. The biggest hurdle for many graduate students is not learning how to conduct research but learning how to write and publish the result from their own work. Early on, their poor writing

---

Box 7.1   How to help (and how not to help) a new graduate student

When I was a first-year graduate student, I signed up for a talk at a conference before I had all my data together or even a real plan of how to present my hypotheses. That isn't necessarily a bad thing, but I was relying heavily on a professor in another department to help me interpret my results. Although that person had been helpful in the data-gathering stage, she was, to my surprise, not helpful during the interpretation phase. In fact, I would go so far as to say that she pretty much left me swinging in the wind when it came time to mentor me. Instead of the help I expected, I was left without a clear way of interpreting or discussing my results. I was distraught because the meetings were coming up in a few weeks. Because this would be my first talk as a Ph.D. student, I was really nervous that I would be making a fool of myself. My primary advisor at the time was aware of all this, but he was letting me figure things out on my own until it was clear I needed some help. He knew I was ambitious, but I was also naïve; I'm glad he saw the value in letting me bite off more than I can chew until he saw that it was time to intervene. When I came to him for help, he sat with me for long hours over the days before the conference and he helped me figure out the best way to discuss my results. Ultimately, with his help, I gave a pretty good talk and I learned a great lesson about being independent (i.e., be independent but know your limits). I also learned that my advisor would have my back when I needed him, and that is something every student should know they have.

---

skills may frustrate you, but try to encourage them to write often and give them the feedback promptly. This is your job, and although you may not want to spend three days editing a poorly written manuscript by a student, it will eventually pay off when you see their improved writing. Be sure to hand them back a revised manuscript where they can follow the changes you've suggested. Go over the changes you've made with them so they can learn from their mistakes. If they make the same mistakes over and over, make it clear to them that they need to fix those errors themselves before you will be willing to review their work further.

Making a New Course

You might love to teach and you may be a fantastic teacher, but if you've never prepared your own course before, you are in for a rude awakening. This isn't like graduate school when you were a teaching assistant running a lab or grading papers and exams. No, you are literally building a course from the ground up. You are making lectures, exams, schedules, assignments, labs, and everything else that goes into writing a new course from scratch. This course prep can be a wonderful experience (no one thinks so when they are actually doing it), but it is certainly extremely time consuming. If research is emphasized at your institution, you should try to teach as few courses as possible before tenure. At Research I institutions, grants and papers are more important than teaching; however, teaching is still an important part of your evaluation. It is usually seen as 50% of your duties as a professor (the other half being research and service). If you negotiate a teaching abstention your first semester, that's great; use that semester to write grants and papers, but every so often start preparing for your new class. If you don't have an abstention, you will have to jump right into teaching, and God help you if you have to prep more than one course at a time. First of all, find out if teaching two classes means you really have to teach two full classes on your own or if one can be team-taught (i.e., you only teach part of the course) or a graduate seminar (e.g., a journal club). Your mentors and advisors will make it clear to you what you are expected to teach, so don't volunteer to do much more than what is expected unless you are at an institution that favors those kinds of teaching initiatives as much as grants and papers. At most Research I institutions, you are there to do research and you should teach classes related to your research. Preparing a new course is an incredible time sink. I would say that for every 10 minutes of actual course time for a new undergraduate class, I spend an hour in prep. That means for a course with three lecture hours a week, I spend nearly 20 hours in preparation. This is even more if there is a lab. Luckily, the students won't be able to tell if you are just one step (one class) ahead of them. It isn't unusual to have no idea what your next lecture will be, particularly by mid-semester. In your first prep for a class, you are basically writing 20- to 30-hour-long talks over the course of a semester. These aren't like writing a research talk where you are basically presenting material that you've been thinking about and working on for a while. To the contrary, in an undergraduate class, you are presenting information on a broad subject to students who are looking to you to explain this difficult information in an understandable way. There is no easy way to prepare a new class, but you can make it slightly easier on yourself with a little foresight. If you are teaching a class that has been taught before, get all of the lectures on that topic that you can, get all of your scientific friends' lectures, and you can even solicit them from people you don't know that have taught the class (they've been in the same boat). The hardest problem is not always the words but the images. Having other people's lectures will save you a tremendous amount of time gathering and organizing the materials you need. However, it is unlikely you want to present the materials just as they did, so you will still have to spend a great deal of time reorganizing slides. If it will help, get a textbook on the subject. This textbook will be your guide to explaining the material. If there are some lecture topics that you can't explain in the allotted time (or that you yourself have difficulty explaining), you can refer the students to the text.

Sometimes the textbook is too much of a crutch or too much of a burden when you realize you want to veer away from a particular subject as it is presented in the text, so be very careful about forcing students to purchase this book. Resentment will grow if you just made them buy a $150 textbook they will never use; instead, offer it as suggested reading and keep a copy in the library if you are not sure how much you will use it. Try to get as many of the lectures written as possible before your course actually starts meeting. Students will want a syllabus and an exam schedule, so as best as you can plot out the major topics that you will discuss before your class starts. It is okay to tell students that this is your first time teaching; they will appreciate your candor and more readily forgive you for mistakes as long as you make efforts to fix them. In fact, before the first exam, give them an informal midterm evaluation form that they can use to grade your performance. By doing this, they can help you implement changes, and they will also appreciate you caring about their opinion. A little effort listening to your students' opinions goes a long way, and you will see their appreciation for your efforts in the official end-of-year teaching evaluations. These official evaluations are included with your tenure package, so take them very seriously.

When dealing with a new class, it is always best to be strict early on and then loosen up as you become more comfortable. It is much harder to go from being a nice teacher to a strict teacher because the students will find it harder to take you seriously based on their first impressions. Imagine being a student in a class who is expecting to have one lecture exam, only to be told mid-semester that there will be three. Students don't like surprises, so let them know as early as possible what they should expect.

You might not have a lot of leeway on what you can teach. Usually, assistant professors are asked to fill some need in the undergraduate curriculum. The nice thing is that you are the boss of any class, so you can steer the class to areas where you are most comfortable. Of course, there is a limit to this; if you are teaching genetics, you probably shouldn't spend three weeks talking about weird diseases without ever talking about Hardy-Weinberg, alleles, punnett squares, and inheritance. If students from your class are leaving without learning the fundamentals, your colleagues will find out, especially if your class is a prerequisite to theirs.

A great way to fill time, that is also a great way to teach, is the use of active learning. There are many books written about active learning that don't need to be introduced here, but basically, active learning turns the regular old lecture-only format into more hands-on and group-thinking exercises. Ultimately, we all want our students to understand the material, and the one-way barrage of information that they get in a lecture isn't necessarily the best way to do that (for you or them). Having students solve problems in class and in groups, or give short lectures themselves, is an easy way to engage students. Even just interrupting your lecture to ask the students direct questions as part of a discussion is a form of active learning.

Teaching a class for the first time can be a harrowing experience, so make sure you are teaching a new class as few times as possible. Teaching the same class for the second time takes less than half the prep time compared with your first round, but make sure you keep good notes on places to improve; otherwise, you'll make the same mistakes all over again. You should also keep students' exams and assignments well after the class is completed. You will be surprised to see students come back years after they've taken the class trying to squeeze a higher letter grade out of you so that they can graduate. Many universities put the burden on you to not only keep this old material but to prove why you gave the grade that you did. It is also good to keep this material in case these students ask for letters of recommendation.

As a time saver and for the sake of grade parity across semesters, you might consider not returning exams to students in a class that you teach multiple times. Old exams inevitably find their way into the hands of the next class of students, giving them an unfair advantage. Another reason to hold on to these exams is to save you time from writing a completely new exam each time you teach the class. If your lectures stay largely the same, most of your exam questions will converge on being very similar if not identical even if you intended on making them different. You must allow students access to their exams so that they can learn from their mistakes, but you might prefer to do this during office hours so that you can ensure that you retrieve the questions. If you go over their exams in an adequate fashion, most students won't mind not receiving the hard copy back.

Service

Part of what you are expected to do as an assistant professor is serve on committees and conduct other service to the department. Although not the most important thing on your pre-tenure resume, doing a good job organizing an event (or a really bad job) will help (or hurt) you when it comes time for the other faulty to vote on your tenure. An intentionally bad job can lead to some faculty turning against you. You may think that service is not important, but service shows that you are a team player and a good colleague.

Some institutions expect their new faculty to do a lot of service. This doesn't mean you have to say yes to everything, but you can't say no to everything either. Do what you can, but avoid committees that are such time constraints that they take away large chunks of time from your research. If possible, serve on committees with deans and chairpersons; it will be in your best interest to get to know these folks better. If you think you would do a terrible job on a committee, be honest with your fellow faculty and volunteer for another committee instead. Find out from past members what the time commitment is for these.

If your institution likes to keep junior faculty off major service detail within the institution, that's certainly great; however, it isn't a bad idea to volunteer for external service outside of your institution anyway. This may include service related to a scientific society, or perhaps being an assistant editor for a journal. You might want to build some clout with your peers at other institutions by organizing a symposium at a national meeting and perhaps editing a book based on those talks. This kind of national or international service builds your reputation within your field and will impress your colleagues at your institution. (Writing a book about the academic process will likely not impress your colleagues, so don't do it. Also I heard there is a really good one on the market right now.) Be careful not to bite off more than you can chew: just because you are elected as the youngest chair of your department doesn't mean that you will get granted tenure for your efforts. In fact, you might be getting yourself into trouble by putting yourself in position to make decisions about salaries and internal grants that may be unpopular among your colleagues. Always be cognizant of your role as a junior colleague; stay low key during faculty meetings unless you have some real insights on a specific topic. You want to make a good impression at these meetings, and unless you have the experience to speak up on a subject, you should not. Your opinions about the future directions of the department don't hold much weight until you convince your colleagues that you will be there in that future. That doesn't mean you should hold your tongue if you feel strongly about something, but you should recognize that you have less clout and fewer bullets in the chamber then tenured faculty (i.e., pick your battles).

Getting Your First Grant

By far, the biggest burden and the most pressure you will feel pre-tenure is getting your first big grant. Each tick of the tenure clock before you have a grant is like the sound of an alarm going off—a reminder of your burden. Remember that great start-up package you got? Well, the department, like a mafia boss, wants that money back with interest, and the only way to appease them is with grants. The earlier you get a grant, the easier time you will have concentrating on publications, teaching, etc. If you are lucky enough to get a grant early that doesn't mean you should stop trying to get your second grant. At many institutions, that second grant is the one that puts you over the top to prove that you are worthy of tenure. The most important thing is to have an active grant at the time that you are going up for tenure.

Unlike papers, you can't submit a grant anytime you want; there are strict deadlines and you should always be mindful of these. Get a strategy in your head about which grants and deadlines you plan on applying for and when you need to start writing them up. A new grant can take months to write, and you might need preliminary data and letters of support that take weeks to gather. Get a calendar and start thinking about what you can apply for in terms of national grants and local grants. Peruse your fellow faculty members' CVs to see what kind of granting agencies they've received funding from. Ask them for copies of their successful grants. Most important of all: start writing. If you're at a loss for ideas, start looking at your own work. Was your Ph.D. thesis something that you can expand into a new larger project? Can you expand your postdoc project into something bigger with your former PI as a collaborator? Can you turn the data from one of your papers into preliminary data for a larger proposal? The most impressive grant proposals have a ton of preliminary data that prove to the reviewers that the work can be completed. You might have tons of preliminary data on these former projects that you can expand on without a great deal of new effort, but always ask for more money then you need. I've been told that you should always be applying for grants for projects that are nearly done, so that you can use that money to get preliminary data for the next grant. That isn't always the best strategy, but it will keep you thinking ahead.

I would recommend writing two big grants in your first two semesters that you have cycling through major federal funding agencies (e.g., NSF, NIH). If there are two cycles each year for grants with a particular agency, you should have something in each round; like the lottery, you have to be in it to win it. Even if there is a slim chance you will be awarded the first time, you can at least find out from the reviews how you should improve for the next round. It is good to show your department that you are doing your best to get funded. They may forgive someone for not having a large grant if they show they've made all attempts to get funded. Someone without money who barely even tried to get a grant will be looked at more negatively. Getting a grant proves that your work is looked at favorably outside of your home institution. Without evidence that your peers outside your institution believe in your work, why would those at your home institution look at your work positively? Your peers understand that the funding rates for major grants are generally less than 10% (sometimes less than 5%). But you don't want to get into a situation where they have to decide between you and rolling the dice on a new and exciting hot-shot candidate they can replace you with. Also, most of your peers with tenure had to get a grant to get tenure, so why would they soften their standards for you?

There might only be one or two times a year when you can apply for a major NSF or NIH grant. In fact, these days you often don't get the reviews back from your January submission until after the July submission deadline (in the case of the two annual rounds for major grants at the NSF). That means you might only have three or four shots to get a particular grant funded before you go up for tenure. If that is the case, you should prepare a grant that you submit in January and another one that you submit in

July (if those are the yearly deadlines). If you have two grants that you can submit, it is against your best interests to submit them to the same panel in the same cycle where they will compete against each other. If you don't get either funded, then you will have to wait an entire year to resubmit them. It is a better strategy to apply to one in January and save the other for July so that you always have a horse in the race in each cycle. If only pre-proposals are requested I would consider submitting more than one at a time, to see which project the granting agency views as the strongest.

To be a slam dunk for tenure, you might shoot for trying to get a grant in all major aspects of your responsibilities. These are typically, in order of importance, (1) research, (2) teaching, and (3) service. It's probably better to get two #1s rather than any of the others if you are at a research I institution, but getting the other two will certainly still look very good if you have a research grant already. Shoot not only for large single PI grants but collaborative grants and smaller exploratory grants. Having funds from several agencies shows that you are well rounded. Even in academic jobs where research is not the emphasis and where perhaps grants are not as important for getting tenure, administrators will still view grants as a major bonus. The administrators who wouldn't know a pipette from a sledgehammer can sure tell the difference between someone bringing in no overhead money and someone bringing in a million dollars. Overhead (or indirect) funds, the grant money that the institution keeps for processing and handling your grant, is something that administrators love because cumulatively it shows how successful a department is at generating money for the larger institution. Remember if you write a collaborative grant that you will only get part of the credit and only part of the overhead. If you are the sole PI on a $500K grant, all that overhead comes to your institution. If that $500K is split among five co-PIs at different institutions, then your grant will be much less impressive to administrators and your colleagues. Your goal should be to have an independent lab where you are funding your own research and your own employees and students. Also keep in mind that, although grants for teaching and renovations are great, they often do not contribute any overhead. If you are at a research I institution, the only thing that will impress your colleagues is your ability to prove yourself as one of the few who can get those 5–10% of the grants that are funded each cycle by NSF or NIH. (There may be other comparable granting agencies for your particular discipline, but NSF and NIH remain the gold standard for most scientists.) You should always be thinking about future projects and what you can get funded—be sure to aim high. Grants aren't given out as a favor to you; money is doled out to exciting projects that will provide the most bang for the agencies' buck. You have to be at the top of your game to get a grant, and you have to distinguish yourself from the 95% of applicants that will be denied funding once again. To be the 5% cream-of-the-crop that get funded, you will need to have a transformative project idea explained in a well-written proposal. You will need preliminary results that act as a proof of concept and cutting edge methods that are pushing the boundaries of your field. You will need to explain why your project is so great and how the results can make an impact on your field and to society (remember those broader impact statements).

Part of being a successful academic is proving that you are innovative and productive, not only by the standards of your peers at your institution, but also to outside reviewers and granting agencies. Your academic success is largely measured by your ability to maintain a well-funded independent research program. Getting grants is the most important part of that success.

Getting Papers Out

Aside from grants, you need to spit out papers as an assistant professor at a steady and impressive pace. A grant alone is not enough. If you get a grant and have zero publications, the chances are near zero that you will ever get another grant because you haven't demonstrated to the granting agency that you can get results. You will need to establish your independent research program and show your colleagues that you can come up with original research and communicate your results to others. See the section above called "The First Year" for more information about publishing pre-tenure. You also would like to go up for tenure with not only publications that are already out, but also publications in press and in review to show that you aren't only concerned with tenure but also with having an active research program for years to come. You should also note that some institutions ask what you have done since tenure when you come up for full professorship. The highest level you can attain at a university is full professor, and typically it is expected that you have roughly 50 publications (with an H-index around 15 or 20) and a history of grants and graduating students (the numbers are all dependent on the institution). It may be to your advantage to think ahead to this full professor stage even before you go up for tenure. Is your goal to one day be a full professor? Is this job the one you want? Is this what you really expected from your institution and the tenure track? People who get tenure are people who really want it; some people who want tenure, don't get it. The only certainty is that no one will get tenure without a great deal of effort and desire. If publishing is not for you and if you don't get pumped up by your research, your colleagues will be able to tell. The advantages of tenure are many and are not limited to job security and the freedom to publish on controversial topics without fear of being fired. Your colleagues will want to know that the person they are giving tenure will continue being an active member of their department, continuing to publish and getting grants. If they don't see that, they will vote to replace you by voting against you during your review. The self-motivation I mentioned in the first chapter for the undergraduate student has to be maintained throughout your academic career, and that hunger in your belly has to be with you throughout.

Mentoring Committees, Reviews

It is likely that your department will set up a mentoring committee of a small group of your peers to help spell out what it is expected of you to get tenure. You will likely meet with this mentoring committee at least once a year to update and prepare your tenure package. These colleagues will evaluate your updated CV, teaching resume, your statements on teaching and research, and your record of getting funding. Generally, they are nice to you early on, but if it is getting close to the time that you go up for tenure and you don't have a grant or enough papers, they will hopefully be honest with you and remind you of the importance of these items. This will be the face-to-face committee that will give you a sense of how the rest of the department will be evaluating you behind your back. If you have a choice of committee members, make sure they are diverse members of the department who will be frank in evaluating your progress. One of these mentors will likely be called upon to present a summary presentation of your tenure package to the rest of the department during the faculty meeting for your tenure vote. (For an example of a tenure-package summary presentation, see Appendix 11.) This mentoring committee isn't like a graduate student committee, they won't be asking you why the sky's blue or how hot the sun is, they will be looking at your CV and evaluating you as a peer in the department based on your record, not your potential. You were hired for your potential; now is the time to show some productivity.

Hopefully, aside from the official capacity of the mentoring committee, you should have some friends in the department whom you can ask for career advice. It is probably a good idea to have coffee, tea, or your favorite adult beverage once a week with someone in the department just to keep up with the gossip and goings-on in the department. As a junior faculty member, you are often left out of the loop, so having other professors to chat with not only keeps you up-to-date but will show that you are likely a sociable and likable colleague.

Tenure Package and Reviews

The requirements for your tenure package are different at every institution, and you should become familiar with these requirements as soon as you can. Most places attempt to make it fairly straightforward; you submit (1) an updated CV in some standard format, (2) your teaching evaluations, (3) statement of research interests (4) statement of teaching philosophy, and (5) a copy of all publications since you joined your institution. Number 5 and the amount of grant money you've brought in are typically the major concerns at a top research institution. The department may also ask your peers in your field to write letters on your behalf, but this is largely out of your hands. Because of this external review, try to keep from burning too many bridges and causing controversy in your field: unless of course it is absolutely necessary. Everyone should have a satisfying list of both professional friends and enemies.

Your tenure talk is the last of the three major talks you will give in your career. The first was your Ph.D. exit seminar, second was your job talk, and now the granddaddy of them all—tenure review. This talk has some flexibility in format. It does not need to be a straight-up research talk that lasts 45 minutes. It should mostly be a research talk discussing aspects of your research program at that institution, but it can also include a list of your grants and students. You can even have a slide or two about teaching (depending on how important that is at your institution). You can talk briefly about renovations or other products that you've contributed positively toward improving your institution. If there was some controversial thing that you did, this is not the time to explain what happened; this will only remind everyone of the fact. Ninety percent of your talk should be about your research program, so don't harp on the peripherals. There will be administrators and peer faculty who may not have seen you since your job interview. Try to make your research understandable at all levels, avoiding obscure acronyms and terms that will go over the heads of some of the audience. Try to talk about the big picture of what you've done. It is also good to talk about the big picture of what you still plan to do—what direction is your lab heading and what does the future hold.

By the time you go up for tenure, you should have a pretty good understanding of how well you will do. Hopefully you've had some pre-tenure reviews where you got some feedback about your progress. Once your tenure package is in and you've given your tenure review talk, there isn't much else you can do. First, you go up for a vote among the tenured faculty in your department. If you pass that, the recommendation for tenure goes to the higher-ups in your department. It might take up to a year from when you submit your package to the time you hear a positive review. Negative reviews will come faster, especially if your peers vote you down. Every institution has their appeal procedures, and this may grant you an extra year, particularly if there was an illness or a pregnancy. But unless you get a major grant or have a slew of papers come out during that time, you are just prolonging the inevitable. Well-liked people sometimes don't get tenure; productive people with grant money sometimes don't get tenure. Sometimes people who don't deserve it get tenure. Just as the case was with job interviews and just as the case is with all review processes, the end is not always just. All you can do is understand what is expected of you and do your best to achieve those goals. Make it your personal goal to have your resume look so good that you will be a no-brainer yes-vote for tenure. The only way to do this is to work very hard and smart. The pre-tenure period should be the time when your research career reaches its apex. Remember, you were hired for your talent and because of your potential. That potential should come into full bloom pre-tenure, and you should be able to build your reputation for your entire career in this period. Post-tenure you can focus more on training people and maintaining a productive and independent research program that will be more about your lab as a whole than about you alone. The pre-tenure period is all about setting up the rest of your career in academia. Success in this period largely is evidence that you will be a successful academic after tenure, too, and that's really what the review is all about.

Summary

- Getting a job is awesome. Congratulations. Remember: Keeping a job is even harder than finding one, so get to work strategizing how you will get grants and papers all while teaching and handling committee work.
- In negotiating start-up, don't just think about your salary and the supplies that you will need. Negotiating things like time off of teaching and a permanent line for a graduate student's salary can be just as important as the monetary components of your start-up.
- Negotiate wisely. Ask the most recent hires what kinds of funds they received and their salary. Ask if there are things they would have done differently and how to best negotiate with the chairperson or dean.
- You know what kinds of equipment and support you will need to get tenure. The people you are negotiating with also want you to get tenure, so be reasonable but also be convincing. Don't walk away from the table without getting the funds you need for essentials and priorities.
- Ask about what kinds of things are expected for you to get tenure. Paying back start-up funds with your own grants and showing that you can run an independent lab with your own federal funding is almost universally important. Having a strong publication record and good teaching evaluations are also important at most places. There shouldn't be any surprises when you go up for review.
- Hire wisely. Think about having a top-heavy lab with postdocs who are gathering data and training students. If you have a bottom-heavy lab with lots of students and no postdocs or technicians, you will be doing all the training yourself along with all the mentoring. As the PI, you should be busy writing papers and grants.
- Try to get the easy papers out first: those side projects you never wrote up from your dissertation, or little ideas you always wanted to try. Getting a nice set of papers out will show you are productive. Spend time early on preparing major grants that will fund your bigger, more ambitious projects.
- Have a big grant written for every major grant cycle and space out the submissions so they are not competing with each other and so that you always have a grant in review and another being revised for the next

cycle. You won't have that many opportunities to get funded, so submit in a strategic way.

- It is better to push yourself really hard early on in your pre-tenure days than trying to play catch-up before you go up for review. If you have ten papers come out at a rate of two a year ever year for five years, that will look a lot better than the person who has ten papers all come out in their last year before review. The pace matters because if it looks like you are just working hard at the last minute to get tenure, your colleagues might think you won't be working very hard after tenure when there is less pressure to be productive.
- Continue to go to conferences and give invited talks. You have to continue to build your reputation. Your department will likely ask people in your field at other institutions to comment on your tenure review. If no one ever heard of you outside of your institution, that will reflect poorly on you and hurt your chances of getting tenure.
- Graduate student training and undergraduate teaching matters. Just because these aspects of your duties are not emphasized as much as research, it isn't license to do a poor job. Administrators especially will find it disturbing if you receive poor evaluations for teaching, and it is often these administrators who have the final say in your tenure decisions and decisions on raises.
- If you are not required to do a heavy load of service for your department, you may look to add national or international service. This looks good not only for tenure but also as part of your synergistic activities for grants. Being an editor of a journal or a symposium organizer can also raise your status among your peers in your field.
- If you are not assigned a mentoring committee of faculty to do a yearly review of your tenure package, it might be worth asking a group of faculty to serve in this capacity informally on your own. You want people to be earnest with you if they think you need to be doing a better job in some aspect of your job. Better to hear it early, while you can still change things.

8 Special Considerations for Women and Minorities, and Balancing Work and Family

The Hard Truth About Being a Minority in Science

Despite many advances in society, women and minorities have additional struggles and difficulties that they must overcome. These surplus problems are the reason for the proportionally few women and minorities we see in academia. Depending on the discipline, the problem can be relatively minor or major, but there is an issue across the board in the sciences. Many books and articles have been written about this issue, so this will be just a brief sojourn (see the works of Joan Williams and Virginia Valian in particular).

As a person who is part of a minority race, but not an underrepresented one, I see the problem of the lack of a proper representation of minorities in academia as a social problem of education at the high school and university level, so the discussion of minorities in science is really beyond the scope of this text. The lack of underrepresented minorities in the sciences and academia in general is a problem of not having enough students of minority races applying to these kinds of jobs, mainly due to a pipeline problem that starts even before the undergraduate level. Many institutions and granting agencies have policies that encourage recruitment and provide grants for minority candidates. Unfortunately, the narrowing of the pipeline starts well before this time. Students and mentors of all races and ethnicities should be aware of conferences like SACNAS (Society for

A Guide to Academia: Getting into and Surviving Grad School, Postdocs, and a Research Job,
First Edition. Prosanta Chakrabarty.
© 2012 John Wiley & Sons, Inc. Published 2012 by John Wiley & Sons, Inc.

Advancement of Chicanos and Native Americans in Science) and HBCU-UP (Historically Black Colleges & Universities – Undergraduate Program) National Research Conference where students of underrepresented minorities can network, be recruited, and learn from mentors who have been in the same situation.

There are internal biases we all have. There have been studies done where the same scientific paper is sent out for review with the author's name being either a "typical" American name, like Matt Davis, or an ethnic name, like say Prosanta Chakrabarty. The same paper authored with the ethnic-sounding name is more critically reviewed than the paper authored with the more familiar name: by reviewers of all backgrounds. You can't do anything about these unfortunate biases; the best way to deal with it is to be aware of it. Only then can you overcome the extra obstacles. If that means making an extra effort to get noticed, so be it. I know that a name like Prosanta Chakrabarty will be difficult for people to spell or remember; for that reason, I've had business cards since I was a graduate student. If you fail in academia, it won't be those little biases that sink you; it will be your own lack of skill, drive, and motivation. If you need to push the accelerator a little harder than the next guy or think a little bit more about how to introduce yourself to people, then, well—that's life. The best thing you can do is remember those struggles when you are at the top of the ladder and try not to have those biases yourself when reviewing and picking candidates. Also, when you are at the top, try to remember to reach back down to help others climb up with you.

The Hard Truth About Being a Woman in Science

A similar but different kind of problem is the lack of women in the sciences. With many Ph.D. programs having nearly as many female candidates as males, the lack of women in science is more of a pipeline problem at the postdoctoral level and beyond. There is a relatively good ratio of men and women in graduate schools, depending on the discipline, but the ratio is terribly skewed at the level of tenure-track faculty in almost all scientific disciplines. Women are also consistently paid lower than their male peers, and any scan of faculty websites will show a dramatic decline of women academics as you head up the pipeline of power. The problem of attrition is usually seen to be most disturbing at the postdoctoral level. It remains unclear exactly why this is the case, but family planning is a likely candidate. There are no easy solutions to these problems. The differences between women and men are not ones of intelligence or desire, only of circumstances and obligations. Women, as the bearers of children, must sometimes choose between the demanding academic schedule and starting or expanding a family. Men must also make this choice, but because they do not become pregnant, they can often take less time off; or at least they are not biologically obligated to take time off. Returning to academia after a substantial break is difficult. Science progresses so quickly that it is hard to get back on the train once you have gotten off. Academia is also a competitive and sometimes ruthless game where rivals can take advantage of those who are absent. No one should have to decide when they can start having a family based on their finances and position in life; the unfortunate truth is that most of us (academics or not) have to make these decisions. Because academics generally spend more time at work than the average person, there is a stigma that academics don't have time for family. I would argue that this is not the case and that academics have greater flexibility in their schedule than most non-academics. The pressures are obviously different between academic and non-academic jobs (you don't have to stand in front of a cash register eight hours a day, but you do have to get your grant submitted by the end of the week), and how you deal with this pressure is more important than how many hours you work.

Academic requirements should never have to hold you back from your personal goals of having and planning a family. I once heard a young female associate professor say, "I could never have gotten to where I am if I had kids." At first that gave me pause, but then I realized she didn't have the experience to make such a statement (i.e., she doesn't know how her life would be different with kids because she didn't have any). It wouldn't be okay for her to say, "If I only had one arm, I would never be able to be president." Only listen to people who have had experience to give you guidance. The biological window for having children is narrower for women than for men, and this constraint may lead to fewer women in academia. The postdoctoral pipeline is often where we see the largest attrition of women. These short-term (1–3 years) academic positions see fewer women than men apply, and we see even fewer women move on from this stage into tenure-track positions.

There is never a perfect time to have a child in academia (or outside of academia for that matter). But given the flexible schedules and consistent pay, I would say an academic job is better than most jobs for someone looking to start or grow a family. The demands of academia may mean having to take home more work than someone who works a 9-to-5 job, but that too can be an advantage: you don't actually have to be at work to get work done. The more troubling practice is taking a year or two away from academia. This absence generally results in a gap in your publication record and CV. Sometimes this is inevitable, but I would say that having a child while in an academic position is better than having a child in-between positions. If you are in a position, taking some time off (a few weeks or a couple of months) usually isn't a big deal (at least compared with many of those 9-to-5 jobs). Your advisor, by the way, has no right to tell you that you don't have time to have a child. You will hear horror stories about female postdocs not having their contracts extended or getting terminated because they become pregnant. There are laws and policies about this kind of discrimination, and you should not stand for any kind of discrimination. Think of the next person who may have to go through the same thing. If you won't fight for yourself, at least stand up for them. Having "that talk" with an advisor about pregnancy should probably only happen once you are actually pregnant. Why mention a pregnancy as a future hypothetical; it isn't their business until the pregnancy is a reality. Honesty works best once there is a reality. If you need to take some time off and if you have doctor's visits that you need to make, remember that your health needs and those of the baby come first. Most people are reasonable, even academic advisors, so be positive and don't worry so much about taking off the time that you need.

Women and men alike should be aware of some "illegal questions" often asked during job interviews. It is not appropriate for someone on a job interview to ask you if you are planning on having children or if you have children. At least not as those questions may relate to your ability to be a viable candidate for that position. These are questions no one can blame you for not answering. You may get asked that question as a matter of small talk, and you can answer it any way you feel comfortable, including by saying you are not comfortable answering. The person may not even be aware of the issues surrounding such a seemingly innocuous question. More serious is when it is clear that you are being judged for your decisions, such as, "I wonder if you will have enough time to work with children on the way." This sort of situation is much less common than it once was, but it does still happen. You never have to apologize or explain your decisions related to your family life in these professional situations. The appropriate way to handle an illegal question might be by saying something like, "I hope you know you are not supposed to be asking me about private matters that don't have anything to do with my work credentials. I will make it clear that my work will continue to be of the highest standard no matter what my decisions are about my private life." If they persist, feel free to say, "That is not a question you should be asking me." If you are asked an illegal question in a way that makes it clear that the person thinks you cannot do top-notch research if you are planning a family, feel free to report this problem to the person running the search or an administrator, discreetly of course ("I don't feel that was an appropriate question to ask me or any candidate").

The discrepancy in women's pay may have more to do with negotiating tactics than sexism, but it is likely both. As I mentioned before, when asked about salary, find out in advance what other people in your position and experience are making and ask for a little more. It might even be to your advantage to say, "Well, I know historically women get paid less than men in the sciences, but I expect the pay to be equitable here. I know your most recent hires make a certain amount and I have the same experience they did when they started." A little quip like this might go a long way to saying that you know your rights, especially if you find it awkward to negotiate your salary.

The Two-Body Problem

You've gone and done it. You fell in love with an academic. Gay or straight, you now have the two-body problem. You want a job in academia and so does your partner. Not only do you need to find a job, but you want to find a job where your partner can also be employed. This isn't as difficult a situation as it once was, but it is still an issue. Sometimes it is impossible to hide the fact that your spouse is an academic. In the age of the internet, it is really hard to keep something like this a secret. One of the other illegal questions you might get during an interview is about your spouse. Obviously you don't say, "I won't come here unless you give my partner a job too." Even if this is true, you are not interviewing together, and they have to choose you before they decide whether they can make accommodations for your partner. Wait until you hold the cards; after you get an offer is the time to ask if accommodations can be made for your spouse. You will be surprised how open some institutions can be. Depending on your partner's position, they may be asked to interview just as you did. In recent years, there have even been federal grants to help institutions accommodate dual hires, such as NSF's ADVANCE: Increasing the Participation and Advancement of Women in Academic Science and Engineering Careers.

The best time to negotiate a dual hire is after you have received an offer and before you formally accept. After you get the phone call that you have been hired, shoot an e-mail to your new chairperson and chair of the search telling them that your significant other is also looking for an academic position and that you were wondering if there were any openings for him/her. Mention that you understand that they might have to go through an interview process just as you did. Depending on the timing, it might turn out that the institution might have an opening for that job description. Perhaps there might not be a full line available at the moment, but sometimes a partial line might be available that will become a full line in the near future. Take what you can get if it makes sense to do so; but it is always the best strategy to ask when you have the power in your hands (i.e., before you accept a pending offer).

Balancing Work and Family

There is no magic formula for how many hours you have to spend at work to be successful, just as there is no magic formula about how much time you are supposed to spend with your family to be a good parent. One hour of interacting with your kids during which they have your complete attention is better than five hours of you being at home but completely ignoring them. It isn't about how much time you spend working or how much time you spend at home; it is about the quality of that time.

When I was a graduate student, I was able to walk to school so I was able to put in long hours and not necessarily in continuous stretches (I could easily go to work from midnight to 5AM if I wanted). When I moved to New York City for my postdoc, I had to commute four hours each day (seriously). There was no way I could work the same number of hours I did in graduate school, unless I planned on not going home at all (not an option). So I learned how to be more efficient out of necessity. (I discovered even more efficient habits after the birth of my twin girls.) Change born out of necessity is what happens to a lot of people when their home life changes either because of a move or because of children. Necessity will dictate how you use every hour of the day, but just because you find yourself with fewer hours for work doesn't necessarily mean your productivity has to suffer. Efficiency will creep in: no more two-hour, two-beer lunches, a little less newspaper browsing, etc. (Read the "Tips on Managing Your Time" section in Chapter 3 for more hints.)

Just because you want to be a great researcher/teacher/scientist/ academic doesn't mean you have to be a bad parent or partner. The academic life is a great one and a flexible one that lets you have room to be good at multiple things. You will have to sacrifice some things to be a great academic, but contrary to popular thought, you don't have to sacrifice everything. The academic-life/home-life balance can be frustrating and draining, but don't use your academic position as an excuse not to fulfill your other wishes for your life. Will your life be more complicated with a new addition to your family? Yes, but it would be in any profession. Just thank your lucky stars that you are in a job that motivates you every day and that allows you to learn and discover new things and provides a flexible schedule. Talk to a 60-year-old academic and then talk to a 60-year-old at nearly any other profession. On average, the academic is still as sharp as he was the day he was hired. He is likely well read outside of professional interests and usually high on the social ladder too—professors, scientists, teachers, academics all. It's a great profession, and one that we are privileged to be a part of.

Appendices

Appendix 1: Example Undergraduate Curriculum Vitae

State University
Somewhere, Louisiana 70800
ann.undergrad@cmail.com
225-555-0000
Website: http://www.ann_undergrad6.net

Education

2007–2011	B.S. in Biological Sciences, Anticipated Graduation: May 2011 (GPA) State University, Somewhere, LA

Research Interests

My ultimate goal is to be a research scientist and professor of biology. I am interested in studying desert life and getting a better understanding of the evolution of adaptations to arid and other extreme conditions.

Research Appointments

2007 – Present	Research Assistant, Museum of Nature, Baton Rouge, LA. Worked under the direction of Dr. Jim Habner Curator of Mammals.
Summer 2009	Research Experience for Undergraduates (REU) student at Sam Houston State University, TX. Worked under the direction of Drs. Chris Randle and Sibyl Bucheli, Professors of Biology.

Scholarships

2009	National Science Foundation, Research Experience for Undergraduates Scholarship at Sam Houston State University ($5000)

A Guide to Academia: Getting into and Surviving Grad School, Postdocs, and a Research Job, First Edition. Prosanta Chakrabarty.
© 2012 John Wiley & Sons, Inc. Published 2012 by John Wiley & Sons, Inc.

2008	Taylor Opportunity Program for Students (TOPS Scholarship, $6000)

Awards and Honors

2007	Einstein Prize for 4.0 GPA in Fall 2007
2007	Valedictorian of Brooklyn High School, Teaneck, LA; Class of 2007

Popular Articles

Undergrad, A. 2011. The undergrad research experience. *LSU Museum Quaterly* (LSU Museum of Natural Science) 29 (1): 15–16. Available online at http://appl003.lsu.edu/natsci/lmns.nsf/$Content/Newsletters? OpenDocument

Research Skills/Experience

(1) Extracted, Amplified (PCR) and Sequenced DNA from Lice and Pocket Gophers - Extractions were conducted using Qiagen Kits; amplification via the Polymerase Chain Reaction and primer sequences for various genes; carried out gel electrophoresis and DNA cycle sequencing, and Sanger sequencing conducted on 3100 sequencer.

(2) Mounted louse specimens – over 1000 mounts produced.

(3) Trapped small mammals (e.g., pocket gophers species, other rodents) in Quintana Roo, Mexico.

(4) Collected scorpions in Tucuman, Argentina and extracted their venom for forensic analysis.

(5) Participated in captive colony care of gopher tortoises in Tuscon, Arizona.

Service

2008 – Present	Member of the National Society for Collegiate Scholars
2008 – Present	Member of Phi Eta Sigma National Honor Society
2008	Attended Women's Leadership Conference, Paris, Texas, July 1
2004	Participated and Placed in Louisiana State Science Fair, introduced interested visitors to photosynthesis using a display and live plants

Important Courses Taken

Zoology, Mammalogy, Organic Chemistry, Calculus, Calculus II, Physics, Ornithology, Molecular Biology, Education in Science
GPA: 3.7; GPA in Major (Biology): 3.9

References

Dr. Prosanta Chakrabarty; Assistant Professor/Curator of Fishes, Louisiana State University, Museum of Natural Science, 119 Foster Hall, Baton Rouge, LA 70803-3216; e-mail: prosanta@lsu.edu; phone: 225-578-3079
Dr. Jim Habner; Professor/Curator of Mammals, Louisiana State University, Museum of Natural Science, 119 Foster Hall, Baton Rouge, LA 70803-3216; e-mail: haf@lsu3.edu; phone: 225-555-3079
Dr. Chris Randle; Professor, Sam Houston State University, 110 Eifel Tower Road, Houston TX, 90210; e-mail: rando@shsu.org; phone: 555-867-5309

Appendix 2: Example Graduate Student CV

Prosanta Chakrabarty
pchakrab@hotmail.edu, 718-647-2192

Education

1996–2000	McGill University, B.Sc. (with Great Distinction) in Applied Zoology, Montréal, Québèc.
2001–current	University of Michigan, Ph.D. Student, Ann Arbor, Michigan.

Publications

Stiassny M.L.J, **P. Chakrabarty**, P. Loiselle. (2001)	Relationships of the Madagascan cichlid genus *Paretroplus*, with a description of a new species from the Betsiboka River drainage of northwestern Madagascar. *Ichthyological Explorations of Freshwaters* 12 (1): 29–40.
Chakrabarty, P. (2001)	Cichlid biogeography: comment and review. *Fish and Fisheries* 5 (2): 97–119.

Published Presentation Abstract

Chakrabarty, P. (2001)	Abstract: Relationships and biogeography of Antillean Cichlids. In: Stevenson, D. W., Abstracts of the 22nd Annual Meeting of the Willi Hennig Society. *Cladistics* 20 (2004): 76–100.

Popular Articles

Chakrabarty, P. (2000)	Faculty of Agriculture and Environmental Sciences. McGill University Guide 2000–2001. p. 20.
Chakrabarty, P. (1998)	The tragic life of Alfred Russell Wallace, *The Afterthought:* MacDonald Campus, McGill University. March 1998, p. 5.

Web Articles

Chakrabarty, P. *Huso huso* (European sturgeon and great sturgeon).
(2001) University of Michigan Museum of Zoology Animal
Diversity Web, http://animaldiversity.ummz.umich.
edu/site/accounts/information/Huso_huso.html

Professional Societies

ASIH American Society of Ichthyologists and Herpetologists, since
1999
SSB Society of Systematic Biologists, since 2000

Field Experience

Winter 1999 *January 4–14, Dominican Republic* I collected freshwater
and marine spp. from 10 localities for morphological and
molecular analyses. The purpose of this trip was to
collect individuals of the two endemic nominal species of
cichlids and discover if they are valid taxa. One of these,
N. vombergae, has not been collected in more than
60 years. Sites visited include the respective type
localities of these two cichlids. Specimens collected were
used in molecular and morphological analyses testing
the validity of these species and in tests of biogeographic
hypotheses.

Research Experience

Summer *NSF funded Research Experience for Undergraduates (REU)*
1999–2000 *student at the American Museum of Natural History in New
York* I described a new species of cichlid from Madagascar,
and synonymized two genera under the supervision of
Curator of Ichthyology, Dr. Melanie Stiassny.
Fall *Research Assistant for the Wildlife Conservation Society, Marine
2000–2001 Program at the Bronx Zoo* I prepared a review and proposal
for the conservation of sturgeon and paddlefish species of
the world. I helped in a campaign to establish a CITES I
listing for the endangered beluga sturgeon (*Huso huso*). I
geo-referenced the Townsend Whale Charts for sperm
whales and northern right whales. I was supervised by the
Director of the Marine Program, Dr. Ellen Pikitch.

Contributed and Invited Talks at Conferences
*Presenter **Poster

Stiassny, M.L.J., Chakrabarty, P.*, Loiselle, P.V. 2000.	Relationships of the Madagascan cichlid genus *Paretroplus* Bleeker 1868, with a description of a new species from the Betsiboka River drainage on northwestern Madagascar. Annual Meeting of the American Society of Ichthyologists and Herpetologists, La Paz, Mexico (14–20 June 2000).
Lauck, E.W.*, Pikitch, E.K., Chakrabarty, P. 2001.	A policy and consumer strategy for sturgeon conservation. Society for Conservation Biology 15th Annual Meeting Hilo, Hawaii, (29 July–August 1, 2001)

Non-Conference Talks

2001	*WCS Sturgeon/Caviar Campaign* to the Wildlife Conservation Society, Marine Program at the Bronx Zoo
2001	*Caviar Emptor* to the Wildlife Conservation Society Marine Program at the Bronx Zoo
2001	*Sturgeon and Paddlefish, Ancient Lineage at Risk of Extinction* to the Wildlife Conservation Society at the Bronx Zoo
2000	*Carboniferous Rhizodont Cranial Fossils; from Horton Bluff* to Vertebrate Paleontology Class of Dr. Robert Carroll
2000	*Saving the St. Lawrence Beluga* presented with Nahanni Fry to the Conservation Biology Class at McGill University

Major Awards & Honors

1999	*McGill University Macdonald Campus Gold Key Award* for outstanding contributions to extra curricular activities
2000	*Valedictorian Class of 2000,* of the Macdonald Campus of McGill University

Grants and Fellowship

1997	*Jean Brown Scholarship* $500, for outstanding contributions to extra curricular activities
1998–2000	*Quebec Bursary for Undergraduates,* covered all tuition costs for undergraduate degree

Teaching Experience

Summer 1993 & 94   Docent at the Aquarium for Wildlife Conservation in Coney Island, Brooklyn, New York.

Workshops/Symposia Attended

Winter 2001   *OSU Phylogenetics Symposium and Workshop*, Ohio State University, Museum of Biological Diversity, December 3–4

Fall 2000   Learning the Sciences: Moving active learning from the lab to the lecture Dr. Chris O'Neal, Instructional Consultant, Center for Research on Learning and Teaching (CRLT), University of Michigan

Contributions/Synergistic Activities

2005   Selected as outside reviewer for submissions to *Journal of Fish Biology, Copeia*.

2004–2005   Led guided tour of the Bronx Zoo for underprivileged children

1996–2000   McGill University, Agriculture, Environmental Sciences Undergraduate Society (AESUS) President, president of over 800 undergraduate students

1998–1999   Teaching Excellence Awards Committee, student representative on committee to select the MacDonald Campus teacher of the year

Research Interests

My research interests stem from my desire to understand fundamental aspects of biological diversity. These fundamental aspects include the relationships of organisms and their morphological complexity. I would like to study these aspects using phylogenetic systematics and geometric morphometrics. These tools will allow me to understand broader themes such as historical biogeography, conservation and the evolution of morphological diversity. I would like to study all three themes by investigating the biology of freshwater fishes. In my biogeographic study I propose using phylogenetic analyses to understand the relationships of Greater Antillean and Central American cichlids. One of the great unknowns in biogeographic studies is the origin of the fauna of the Great Antilles and Central America. By studying the relationships of the fishes on these landmasses we will gain a better understanding of Earth history.

Another great unknown is the relative magnitude of morphological diversity as it is distributed among groups. I would like to use geometric morphometric techniques and analyses of disparity to test hypotheses about cichlid morphological evolution. These hypotheses would be tested by quantifying morphological diversity. I would like to investigate how Rift Lake cichlid groups differ in morphological diversity relative to species richness and ecological diversity. This project would help us get a better understanding on the role of ecology and evolution on morphology.

Appendix 3: Example Thesis Proposal

PHYLOGENETIC AND BIOGEOGRAPHIC ANALYSES OF THE
GREATER ANTILLEAN AND MIDDLE AMERICAN CICHLIDAE

Prosanta Chakrabarty

Department of Ecology and Evolutionary Biology
Museum of Zoology, Fish Division
Ann Arbor, MI 48109–1048

Achieved Candidacy: Winter 2003
Years in PhD Program: 4th year begins Fall 2004
Proposed Date of Completion: 5th year/Winter Term 2006

A dissertation proposal submitted for the yearly meeting of the following
thesis committee:
Dr. William L. Fink (Chairman)
Dr. Gerald R. Smith (EEB member)
Dr. Diarmaid Ó Foighil (EEB member)
Dr. Daniel C. Fisher (Cognate member)

Progress Report

A Guide to Academia: Getting into and Surviving Grad School, Postdocs, and a Research Job,
First Edition. Prosanta Chakrabarty.
© 2012 John Wiley & Sons, Inc. Published 2012 by John Wiley & Sons, Inc.

Progress Report

Activities and accomplishments since last meeting (Winter Term 2003)

Publications

Chakrabarty, P. (2004) Cichlid biogeography: comment and review. Fish and Fisheries 5 (2): 97–119.

Chakrabarty, P. (accepted pending revisions) Testing past conjectures regarding the morphological diversity of rift lake cichlids. *Copeia*

Pikitch, E.K., Doukakis, P., Lauck, L., Chakrabarty, P. (in review) Sturgeon and paddlefish fisheries of the world: conservation strategies for the future. *Fish and Fisheries* (submitted August 2003)

Published Abstracts

Chakrabarty, P. (2004) Abstract: Relationships and biogeography of Antillean Cichlids. In: Stevenson, D. W., Abstracts of the 22nd Annual Meeting of the Willi Hennig Society. *Cladistics* 20 (2004): 76–100.

Conference Presentations

Chakrabarty, P. 2003. Relationships of Antillean cichlids and biogeographic consequences. Annual Meeting of the American Society of Ichthyologists and Herpetologists, Manaus, Brazil (26 June–1 July 2003).

Chakrabarty, P. 2003. Relationships and Biogeography of Antillean Cichlids. Willi Hennig Society; Hennig XXII (20–24 July 2003).

Chakrabarty, P. 2004. Systematics and historical biogeography of the Neotropical cichlid genus *Nandopsis*. Annual Meeting of the American Society of Ichthyologists and Herpetologists, Norman, Oklahoma (26–31 May 2004).

Field Experience

Winter 2004   January 4–14, Dominican Republic, collecting freshwater and marine spp.10 localities, for morphological and molecular analyses

Grants

2003   *Rackham Travel Grant*, $ 900 for travel to ASIH

2003   *EEB Travel Grant*, for the amount of $ 100 for presenting at Hennig XXII

2003   *Rackham Discretionary Fund*, for $ 1250 for research equipment

2004   *Rackham Travel Grant*, $ 400 for travel to ASIH 2004

2004   *Museum of Zoology Hinsdale Scholarship Award* $2,247 for research

2004   *EEB Block Grant/Okkelberg Award* $1,257 for research

Chapter I: CICHLID BIOGEOGRAPHY: COMMENT AND REVIEW

Hypotheses tested:

(1)   The distribution of the Cichlidae reflects a pattern congruent with the break-up of Gondwana.
(2)   Current molecular clock estimates and the known fossil record underestimate the age of Cichlidae.

Phylogenetic analyses dealing with disjunct distributions (distributions that require marine dispersal or vicariant events) are reviewed for the Cichlidae. The most corroborated relationship between clades across a Gondwanan disjunction is the sister relationship between Indian and Malagasy cichlids. The minimum age of the Cichlidae as implied by the fossil record is at odds with the timing of the Cretaceous break of the Indian-Madagascar landmass. All well sampled phylogenies for this group fit a pattern reflecting Gondwanan break-up. Grounds for strictly dispersalist hypotheses are not well founded for any cichlid disjunct distribution, leaving vicariance alternatives as the only explanation.

Status: Published

Chakrabarty, P. 2004. Cichlid biogeography: comment and review. Fish and Fisheries 5: 97–119.

Chapter II: TAXONOMIC REVIEW OF HISPANIOLAN CICHLIDS

Hypotheses tested:

(1)   *Nandopsis vombergae* is a junior synonym of *Nandopsis haitiensis*.
(2)   'Cichlasoma' *woodringi* is a member of the genus *Nandopsis*.

Nandopsis vombergae (Ladiges 1938) is one of three described cichlid species from Hispaniola. This species, known only from its holotype, is a junior synonym of the widespread island species *Nandopsis haitiensis* (Tee-Van 1935). The holotype lacks apomorphies that could be used to discriminate it from *N. haitiensis*. Meristic counts are equivalent, and a Principal Components Analysis (PCA) was unable to discriminate between these two species. The Miocene fossil, 'Cichlasoma' *woodringi* Cockerell 1928 closely resembles *Nandopsis haitiensis*; however, poor preservation precludes an adequate assessment of its status as a distinct species. Diagnostic characters of *Nandopsis* are also described for the first time.

Status: ready for submission to *American Museum Novitates* in 09/04

Chapter III: RELATIONSHIPS AND BIOGEOGRAPHY OF ANTILLEAN CICHLIDS

Hypotheses tested:

(1) The phylogeny of the Greater Antillean cichlids recovers a pattern that is congruent with either
   (a) an Early Oligocene landbridge connecting South America to the Greater Antilles
   (b) a vicariance event between these islands and Middle America in the Mesozoic
   (c) or marine dispersal

Geological reconstructions proposed for the Greater Antilles fit into two major categories, each with different biogeographic consequences. One category of reconstructions suggests that an Early Oligocene landbridge connected the Greater Antillean islands to northern South America (Iturralde-Vinent and MacPhee 1999). The other suggests a period of coalescence in the Mesozoic of some of the Antillean islands with portions of Central America (Rosen 1975). A molecular phylogenetic analysis based on nuclear and mitochondrial genes that includes the endemic cichlid faunas of the Greater Antilles, as well as from South and Central America, results in a phylogeny that is congruent with the later reconstruction.

Status: in prep., data has been collected, presented at ASIH 2003, Hennig 2003

Chapter IV: TOTAL EVIDENCE PHYLOGENETIC ANALYSIS OF MIDDLE AMERICAN CICHLIDS

Hypotheses tested:

(1) The eight sections of 'Cichlasoma' are monophyletic
   C. (*Amphilophus*)'
   'C. (*Archocentrus*)'
   'C. (*Herichthys*)'
   'C. (*Nandopsis*)'
   'C. (*Paraneetroplus*)'
   'C. (*Theraps*)'
   'C. (*Thorichthys*)'
   'C. (*insertae sedis*)'

The majority of the nearly 100 species of Middle American cichlids were placed in the catch-all genus *Cichlasoma* (Regan 1906–1908). When

Figure A3.1   Preliminary phylogenetic analysis of Middle American cichlids. Geographically South American species that are nested within the Middle American group are optimized as Middle American cichlids, making this group monophyletic.

Kullander (1983) restricted *Cichlasoma* to a dozen South American species, the Middle American forms (now recognized as 'Cichlasoma') were left without generic placement. Little work has been done to determine phylogenetic relationships of the Middle American cichlids. Miller (1966, 1976) placed many of the Middle American cichlids into eight sections of 'Cichlasoma'; however, these sections are informal assemblages of species that share similarities in trophic morphology.

A preliminary molecular analysis finds that the Middle American cichlids form a clade that is sister to the Cuban cichlids and that none of the 'Cichlasoma' sections are monophyletic (Fig. A3.1). A phylogeny will be constructed for the Middle American cichlids using mitochondrial genes COI (Folmer et al. 1994) and 16S (Koucher et al. 1989), nuclear genes TMO4C4 (Streelman and Karl 1997), Histone H3 (Colgan et al. 1998), S7 (Chow and Hazama 1998), MN32-2 (Buonaccorsi et al. 1999) and Sactin (McDowell and Graves 2002) as well as over 100 morphological characters from external features, oral jaw ligaments, neurocranial frontal pores, swimbladder extensions and the pharyngeal apophysis.

Status: This chapter is incomplete and will require collection of wild caught specimens and additional molecular work and morphological. See budget and timeline for details.

Time Line for Thesis Completion

Fall 2004: Funding as Teaching Assistant
   Submit Chapter II to *American Museum Novitates*
   Send in revision of possible appendices chapter to *Copeia*
   Reapply for NSF Doctoral Dissertation Improvement Grant
   Apply for Rackham Predoctoral Fellowship Award
   Continue morphological work for Chapter IV

Winter 2005: Funding as Research Assistant
> Possible collection trip to Nicaragua, Mexico or Cuba
> Submit Chapter III for *Molecular Phylogenetics and Evolution*
> Complete molecular work for Chapter IV
> Continue morphological work for Chapter IV

Spring/Summer 2005
> Present 'Phylogeny of Middle American cichlids' at ASIH
> Complete morphological work for Chapter IV

Fall 2005
> Write up Chapter IV and submit
> Write up dissertation and submit
> Apply for NSF Postdoctoral Fellowship

Winter 2006
> Defend dissertation

Possible Appendices Chapters

A   Chakrabarty, P. (accepted pending revisions) Testing past conjectures regarding the morphological diversity of rift lake cichlids. *Copeia*

B   Pikitch, E.K., Doukakis, P., Lauck, L., Chakrabarty, P. (in review) Sturgeon and paddlefish fisheries of the world: conservation strategies for the future. *Fish and Fisheries*

C   Fink, W.L. and Chakrabarty, P. (in prep.) Phylogeny of the piranha genus *Sarrasalmus*.

Budget

Laboratory Work (estimated for 150 taxa)

Quantity Total Price

DNA extraction
Qiagen Dneasy kit (50 columns)
2                              $ 100
1.5 ml microcentrifuge tubes
10 cases $ 130
Micropippette tips ($55 per rack of 1000)
10                            $ 550

Polymerase Chain Reaction
Platinum Taq Polymerase (50 units@ $53.63 ea)
10                            $ 540

DNA purification
QIAQuick gel extraction kit (250 columns)
5                              $ 1,674

Sequencing Fees
Sequencing Core lane fees (8 lanes per taxa @ $3/lane)
1200 $ 3,600

Purchasing of Voucher specimens
$ 10 average per specimen (50 to 70 specimens, plus shipping)
$ 400

Conferences
ASIH 2005 (Tampa, Florida)
Evolution/SSB 2005 (University of Alaska, Fairbanks)
ASIH 2006 (New Orleans, Louisiana)
                                   Cost: $2000

Potential Field Work
Cuba (Transport, lodging, food)
Nicaragua (Transport, lodging, food)
Mexico (Transport, lodging, food)
Cost per trip: $ 2000

                    TOTAL $10,994

Sources of Funding for Research:
Rackham Block Grant, Rackham Discretionary Funds, Hinsdale-Walker
    Grants, National Science Foundation Dissertation Improvement Grant,
    American Society of Ichthyology and Herpetology Gaige Award

Appendix 4: Example Job Cover Letter

M U S E U M   S T A T E   U N I V E R S I T Y

January 1, 2011

Dear Selection's Committee:

Mu.S.U

As a museum trained ichthyologist and evolutionary biologist I am excited to apply for a position where I can combine traditional collections based research with cutting-edge evolutionary biology in an environment that encourages both the training of graduate students and course-based teaching. I have been in my current position as a postdoctoral fellow at MuSU for just under two years. In those two years I have shown that I am among the best young ichthyologists in the country. I am well funded (100K + in NSF grants) and very productive (six peer reviewed pubs in 2010, six others submitted). In my tenure at MuSU I have conducted fieldwork in Australia, Vietnam, Taiwan and the Gulf of Mexico. I conduct research in both marine and freshwater systems focusing on systematic studies related to historical biogeography, and the evolution of bioluminescence and taxonomy. I believe my work over the past two years showcases my abilities as an independent researcher. If I were lucky enough to be considered for the American Institute of Zoology position I would be able to expand my research program and take advantage of the many opportunities available to someone working at one of the country's best universities.

Having been trained in ichthyology at MuSU and the University of Michigan I understand how large, collections-based research universities work. I understand that being a researcher at the AIZ comes with the responsibility of being a professor, collector, public figure, and grant recipient all while standing with the top tier of biologists in the country. I hope that I can convince you that I am a worthy candidate to become your colleague.

A Guide to Academia: Getting into and Surviving Grad School, Postdocs, and a Research Job, First Edition. Prosanta Chakrabarty.
© 2012 John Wiley & Sons, Inc. Published 2012 by John Wiley & Sons, Inc.

Within this application I have included my curriculum vitae, three publications, a research statement and a curatorial statement. I have included this curatorial statement because I think curation will be an important aspect of this new position. In that statement I discuss not only my recent work cataloging and creating a molecular collection but also ideas for using collections for promoting research and teaching, as well for fundraising through exhibits. I also hope to have the opportunity to show you how I can use my experiences as a biologist to solve issues not only concerning systematics and evolution but also social issues, conservation and broader themes. At the AIZ, I will strive to be successful as a scientist in my field but I also will use my expertise to contribute to the larger community and to solve issues that affect us all. I look forward to teaching in your biology department and my teaching statement describes some of the classes I think will be a great addition to your curriculum.

If given the opportunity to present some of my research to the other scientists at the AIZ and adjacent units, I will show myself to be a capable ichthyologist motivated to solving some difficult biological questions. This important position would allow me the chance to work with many wonderful and talented people. I think the tools that I have in phylogenetics, molecular biology, and geometric morphometrics will allow me many opportunities to collaborate with others at the AIZ. I think I would be an excellent complement to the outstanding group of scientists currently at your institution; I look forward to meeting you.

Sincerely,

Prosanta Chakrabarty
Postdoctoral Fellow
Museum State University
225-555-3079
prosanta@musu.edu.org

References

(1) Pelanie M.J. Shiassny; Curator, American Museum of Natural History; Central Park West at 345th Street, NY, NY 10029. pmjs@amnh.edu, 1-212-555-5796.
(2) Billiam L. Mink; Curator, University of Michigan Museum of Zoology; 1111 Gedders Ave., Ann Arbor, MI 48105, bmink@umich.org, 1-734-555-9928.

(3)   Donn X. Starks; Curator, American Museum of Natural History; Central Park West at 345th Street, NY, NY 10029. dxstarks@amnh.edu, 1-212-555-7791.

Please feel free to contact anyone at my previous or current departments as references on my behalf. All of my publications and additional information about my research program are available on my website (http://www.prosanta.net).

Appendix 5: Example Research Interests for Job Application

SUMMARY OF RESEARCH INTERESTS

Prosanta Chakrabarty

My current research interests stem from my desire to understand fundamental aspects of biological diversity. These fundamental aspects include the relationships of organisms and their morphological complexity. I study these aspects using phylogenetic systematics, geometric morphometrics and other tools. These tools allow us to understand broader themes such as historical biogeography, molecular evolution, conservation and the evolution of morphological diversity. As an ichthyologist, my focus is on process oriented evolutionary biology in marine fishes and pattern oriented evolutionary biology in freshwater fishes. As a systematist, I use molecular and morphological tools to help discover relationships among species and resolve taxonomy in order to better understand the evolutionary history of fishes.

My most recent research focuses on understanding the phylogenetic and biogeographic relationships of freshwater fishes from Middle America (Central America, Mexico, Greater Antillean Islands) and on the phylogenetic and evolutionary history of deep-sea fishes in the Gulf of Mexico and the Indo-West Pacific. I am also conducting studies of blind cave fishes of Gondwanan origin and bioluminescent fishes from the deep ocean.

Systematics of Deep-Sea Fishes

Part of my current research focus is to resolve some of the difficult phylogenetic problems associated with bioluminescent and deep-sea fishes. In recent years I have focused on the Indo-Pacific family of ponyfishes (Leiognathidae; with collaborator Donn Starks, AMNH) and on deep-sea forms including batfishes and tripod fishes (with Hans Ho of the Academia

---

A Guide to Academia: Getting into and Surviving Grad School, Postdocs, and a Research Job,
First Edition. Prosanta Chakrabarty.
© 2012 John Wiley & Sons, Inc. Published 2012 by John Wiley & Sons, Inc.

Sinica and Matthew Davis of the University of Kansas). Many deep-sea fishes are bioluminescent and co-opt the light produced by bacteria for sexually dimorphic photic communication. By resolving the relationships among these fishes we can better understand how this ability may have influenced their evolutionary history (e.g. increased speciation rates or morphological disparity). I am also currently resolving the many taxonomic problems that plague these groups using molecular, morphological and behavioral (flashing) characters in phylogenetic analyses. Much of this work is associated with the products of my recent field trips to Vietnam, Madagascar, Sri Lanka, Indonesia, the Malay Peninsula and Taiwan. The phylogenies being prepared for these families will be used to revise the taxonomy of these group, identify new species, and test hypotheses about the role of sexual selection and photic communication on morphological diversity and species richness.

Recent Publications on Deep-Sea Fish Systematics:

Ho, H.C., Chakrabarty, P., Sparks, J.S.(2010) Review of the *Halieutichthys aculeatus* species complex (Lophiiformes: Ogcocephalidae), with descriptions of two new species. *Journal of Fish Biology* 77:841–869.

Davis, M.P., Chakrabarty, P. (in press) Video Observation of a Tripodfish (Aulopiformes: *Bathypterois*) from Bathypelagic Depths in the Campos Basin of Brazil. *Marine Biology Research*.

Pending Deep-Sea Grants:

$ 601,414* submitted   NSF DEB: The Timing and Evolutionary History of Deep-Sea Invasions and Habitat Shifts: Systematics and diversification within lizardfishes, cods and batfishes. PC is CO-PI, PI is Matthew P. Davis (LSU) *submitted July 9, 2010.

Historical Biogeography

I am also currently interested in the historical biogeography of Neotropical fishes. I hope to better understand the biogeographic relationships of Greater Antillean, Middle and South American fishes. One of the great unknowns in biogeographic studies is the origin of the fauna of the Greater Antillean islands and the four different geological blocks that have united to form Central America. I found evidence through phylogenetic analysis that the origin of some of the freshwater ichthyofauna is linked to ancient vicariance events. Vicariance was favored over alternatives because both the pattern of the recovered phylogeny and divergence time estimates were congruent with this hypothesis. I continue testing these hypotheses and several large-scale phylogenies that include a much broader sampling of taxa including fossils. The relationships of each of these groups are unique but all of them can potentially test biogeographic patterns for the region through phylogenetic analyses.

Recent Biogeography Publications:

Smith, W.L., Chakrabarty, P., Sparks, J.S., (2008) Phylogeny, taxonomy and evolution of Neotropical cichlids (Percomorpha: Cichlidae: Cichlinae). *Cladistics* 24, 1–17.

Chakrabarty, P. Albert, J. (in press.) Not so Fast: Freshwater Fishes and the Great American Biotic Interchange. In: *Historical Biogeography of Neotropical Freshwater Fishes* (J. S. Albert and R.E. Reis, eds.)

Current Biogeography Grants:

$ 519,944[AWARDED]   NSF REVSYS: Reconstructing Heroini (Teleostei:Cichlidae) – Of Heroes, Convicts, Angels and Red Devils. PC is sole PI. September 2009–2012

Geometric Morphometrics And Analyses of Disparity

I have done a number of projects measuring and comparing morphological diversity as it is distributed among groups. I used geometric morphometric techniques and analyses of disparity to test hypotheses about morphological evolution in several fish clades. Hypotheses about the origins of various groups can often be tested by quantifying morphological diversity. I have published on how African Rift Lake cichlid taxa differ in morphological diversity relative to species richness and ecological diversity. Researchers in the past have tied the rich diversity of these cichlids to different evolutionary hypotheses (e.g., microallopatric speciation, sympatric speciation). Some of these evolutionary hypotheses were based on assumptions made about the morphological diversity of some cichlid clades. I quantified morphological diversity to test these hypotheses. I have also used geometric morphometric tools to clarify taxonomy and to test the role of sexual selection on the morphological evolution of bioluminescent fishes.

Recent Geometric Morphometric Publications:

Chakrabarty, P., Chu, J., Luthfun, N., and Sparks, J.S. (2010) Geometric morphometrics uncovers an undescribed ponyfish (Teleostei: Leiognathidae: *Equulites*) with a note on the taxonomic status of *Equula berbis* Valenciennes. *Zootaxa* 2427:15–24.

Chakrabarty, P., Davis, M.P., Smith, W.L., Baldwin, Z.H., Sparks, J.S. (submitted) Is sexual selection driving diversification of the bioluminescent ponyfishes (Teleostei: Leiognathidae)? *Molecular Ecology* (September 24, 2010).

Taxonomy and Morphological Studies

Morphological character analyses are essential in phylogenetics and descriptive work to ensure proper diagnoses, taxonomy and rigorous sampling of all phylogenetically informative material. As a collections-based scientist I will always continue to have morphological systematics as a major part of my research program. Phylogenies that include morphological characters lead to the discovery of morphological synapomorphies that can be used in descriptive works diagnosing novel taxa or higher groups. Another important use of morphological work is in the use of fossil calibrations in molecular studies. Calibrations based on fossil taxa are widely used to age different nodes in molecular phylogenies. Unfortunately, these fossils are often incorrectly placed on the molecular phylogeny because of the lack of any morphological data from extant taxa in the character matrix. When fossils are available I often incorporate them into my analyses and sometimes redescribe them, as I did with *Nandopsis woodringi* (Chakrabarty, 2006-Taxonomic status of the Hispaniolan Cichlidae). Taxonomy is an important part of systematics and it is a study that collections-based scientists should view as an important part of their research program.

Recent Taxonomy and Morphological Publications:

Chakrabarty, P., (2010) Status and phylogeny of Milyeringidae, with the description of a new blind cave fish from Australia, *Milyeringa brooksi*, n. sp. *Zootaxa* 2557:19–28.

Schaefer, S.A, Chakrabarty, P., Geneva, A., Sabaj-Perez, M.H. (in press) Nucleotide sequence data confirm monophyly and local endemism of variable morphospecies of Andean astroblepid catfishes (Siluriformes: Astroblepidae). *Zool. J. Linn. Soc.*

Philosophy

As a scientist I sometimes think it is important to comment on philosophical matters related to biology and to articulate justifications for one side of a particular scientific debate. Recently, I've found a number of different outlets to make philosophical comments. In one example, I found that molecular phylogenetics and taxonomy were working independently of each other and proposed a new approach. This approach introduces a new nomenclature to molecular taxonomy called "genetypes." Genetypes are sequence data from type specimens. Being able to identify sequence data from type specimens (particularly topotypes) creates a new dimension by which taxa can be compared. In a second example, I've commented on what is seen by some traditionalists as a paradigm shift in systematic ichthyology. This shift is caused by the recent trend in ichthyology to change higher-level fish taxonomy based on molecular data. I think the incorporation of molecular data into this aspect of systematic ichthyology is a positive step and provides a shift from the authority-based taxonomy that dominated the field in the past.

Recent Philosophy Publications:

Chakrabarty, P., (2010) Genetypes: a concept to help integrate molecular systematics and traditional taxonomy. *Zootaxa* 2632: 67–68.

Chakrabarty, P. (2010) The transitioning state of systematic ichthyology. *Copeia* 2010 (3):513–514.

Appendix 6: Example C.V. for Job Application

Prosanta Chakrabarty
Curriculum Vitae

Museum State University

Phone: 225-578-3079

115 Rooster Hall

Fax: 225-578-3075

Department of Ichthyology

e-mail: prosanta@lsu.edu

Cotton Lane, LA 70816 USA

http://www.prosanta.net

SEE COMPLETE CV AT WWW.PROSANTA.NET

Education

2001–2006	University of Michigan, Ph.D. in Ecology and Evolutionary Biology, Ann Arbor, Michigan
1996–2000	McGill University, B.Sc. (with Great Distinction) in Applied Zoology, Montréal, Québec

Research and Academic Appointments

2006–2008	Post-doctoral Fellow, Museum State University, Department of Ichthyology
2001–2006	Ph.D. Student, University of Michigan Museum of Zoology
2000–2001	Research Assistant, Wildlife Conservation Society, Bronx Zoo

Peer-Reviewed Publications

(16) **Chakrabarty, P.**, Davis, M.P., Smith, W.L., Baldwin, Z.H., Sparks, J.S. (submitted) Is sexual selection driving diversification of the bioluminescent ponyfishes (Teleostei: Leiognathidae)? *Molecular Ecology* (September 24, 2010)

(15) Wiley, E.O., **Chakrabarty, P.**, Craig, M.T., Davis, M.P., Holcroft, N.I., Mayden, R.L., Smith, W.L. (accepted) Will the real phylogeneticists please stand up? *Zootaxa*. (submitted May 28) – [INVITED PAPER]

A Guide to Academia: Getting into and Surviving Grad School, Postdocs, and a Research Job,
First Edition. Prosanta Chakrabarty.
© 2012 John Wiley & Sons, Inc. Published 2012 by John Wiley & Sons, Inc.

(14) **Chakrabarty, P.**, Davis, M.P., Berquist, R., Gledhill, K., Sparks, J., and Frank, L. (in review) In situ description of the bioluminescent light organ of ponyfishes using MRI with a phylogenetic review of their photic signals. *Journal of Morphology* (accepted pending revision May 10, 2010)

(13) Davis, M.P., **Chakrabarty, P.** (in press) Video Observation of a Tripod-fish (Aulopiformes: *Bathypterois*) from Bathypelagic Depths in the Campos Basin of Brazil. *Marine Biology Research* (March 26).

(12) Schaefer, S.A, **Chakrabarty, P.**, Geneva, A., Sabaj-Perez, M.H. (in press) Nucleotide sequence data confirm monophyly and local endemism of variable morphospecies of Andean astroblepid catfishes (Siluriformes: Astroblepidae). *Zool. J. Linn. Soc.*

(11) **Chakrabarty, P.** Albert, J. (in press) Not so Fast: Freshwater Fishes and the Great American Biotic Interchange. in: Historical Biogeography of Neotropical Freshwater Fishes (J. S. Albert and R.E. Reis, eds.) – [INVITED PAPER]

(10) **Chakrabarty, P.** (2010) Genetypes: a concept to help integrate molecular systematics and traditional taxonomy. *Zootaxa* 2632: 67–68.

 (9) **Chakrabarty, P.** (2010) The transitioning state of systematic ichthyology. *Copeia* 2010 (3):513–514.

 (8) Ho, H.C. **Chakrabarty, P.**, Sparks, J.S., (2010) Review of the *Halieutichthys aculeatus* species complex (Lophiiformes: Ogcocephalidae), with descriptions of two new species. *Journal of Fish Biology* 77: 841–869.

 (7) **Chakrabarty, P.** (2010) Status and phylogeny of Milyeringidae, with the description of a new blind cave fish from Australia, *Milyeringa brooksi*, n. sp. *Zootaxa* 2557: 19–28.

 (6) **Chakrabarty, P.**, Ho, H.C., Sparks, J.S. (2010) Review of the ponyfishes (Perciformes: Leiognathidae) of Taiwan. *Marine Biodiversity* 140:107–121.

 (5) **Chakrabarty, P.**, Chu, J., Luthfun, N., and Sparks, J.S. (2010) Geometric morphometrics uncovers an undescribed ponyfish (Teleostei: Leiognathidae: *Equulites*) with a note on the taxonomic status of *Equula berbis* Valenciennes. *Zootaxa* 2427: 15–24.

 (4) Smith, W.L., **Chakrabarty, P.**, Sparks, J.S., (2008) Phylogeny, taxonomy and evolution of Neotropical cichlids (Percomorpha: Cichlidae: Cichlinae). *Cladistics* 24, 1–17.

 (3) **Chakrabarty, P.**, Sparks, J.S. (2008) Diagnoses for *Leiognathus* Lacepède 1802, *Equula* Cuvier 1815, *Equulites* Fowler 1904, *Eubleekeria* Fowler 1904, and a new ponyfish genus (Teleostei: Leiognathidae). *American Museum Novitates* 3623: 1–11.

 (2) **Chakrabarty, P.**, Sparks, J.S. (2007) Phylogeny and taxonomic revision of *Nuchequula* Whitley 1932 (Teleostei: Leiognathidae), with the description of a new species. *American Museum Novitates* 3588: 1–28.

(1) Sparks, J.S., **Chakrabarty, P.** (2007) A new species of ponyfish (Teleostei:Leiognathidae: *Photoplagios*) from the Philippines. *Copeia* 2007 (3): 622–629.

Other Publications/Programs

Programs

(1) Janies, D., Hardman, J., Lam, C., **Chakrabarty, P.** (2010) *DepthMap*. This program is a web-based application for recreating baseline distribution records for wildlife affected by the 2010 Gulf of Mexico Oil Spill. (available on-line at http://www.depthmap.osu.edu/)

Book Reviews

(2) Chakrabarty, P. (2009) [Review of] Your Inner Fish, A Journey into the 3.5-Billion-Year History of the Human Body. *Copeia* 2009 (2), 421.

(1) Chakrabarty, P. (2007) [Review of] An Atlas of Michigan Fishes with keys and illustrations for their identification. Bailey, R.M., Latta, W.C., Smith, G.R. *Copeia* 2007 (1): 238–239.

Published Abstracts

(2) Chakrabarty, P., Smith, W.L., Sparks, J.S. (2010) Abstract: Testing the role of sexual selection in the evolution of ponyfishes. In: Szumik, C., Goloboff, P. A summit of cladistics: abstracts of the 27th Annual Meeting of the Willi Hennig Society and VIII Reunion Argentina de Cladistica y Biogeografía. *Cladistics* 26: 202–226.

(1) Chakrabarty, P. (2004) Abstract: Relationships and biogeography of Antillean Cichlids. In: Stevenson, D. W., Abstracts of the 22nd Annual Meeting of the Willi Hennig Society. *Cladistics* 20 (2004): 76–100.

Popular Articles/Letters

(2) Chakrabarty, P. (2008) *AIR* Vents: Chakrabarty, Fish Photo Detective. *Annals of Improbable Research* 14 (3): 3.

(1) Chakrabarty, P. (2006) The Greater Antillean Cichlids. *Buntbarsche Bulletin* 234: 18–22.

Web Articles

(2) Chakrabarty, P. (2007) The fangtooth: *Anoplogaster cornuta*. *Digital Fish Library*, http://www.digitalfishlibrary.org

(1) Chakrabarty, P. (2003) *Huso huso* (European sturgeon and great sturgeon). University of Michigan Museum of Zoology *Animal Diversity Web*, http://animaldiversity.ummz.umich.edu/site/accounts/information/Huso_huso.html

Profiles/Media

ScienceDaily. (2010) Supercomputers Help Track Species Affected by Gulf Oil Spill. August 25 http://www.sciencedaily.com/releases/2010/08/100825111359.htm [Picked up by over 1200 news outlets from EurekAlert]

Shogren, Elizabeth. (2010) Scientists: Dispersants Compounded Oil Spill. National Public Radio, June 7 – http://www.npr.org/templates/story/story.php?storyId=127525694

Grants and Fellowships, Other Funds
Louisiana State University

$ 128,775* submitted	Board Of Regents, Traditional Enhancement Grant – Education: Making a Big Splash with Louisiana Fishes: A Three-tiered Education Program and Museum Exhibit. PC is CO-PI, PI is Dr. Sophie Warny (LSU), October 9, 2010.
$ 601,414* submitted	NSF DEB: The Timing and Evolutionary History of Deep-Sea Invasions and Habitat Shifts: Systematics and diversification within lizardfishes, cods and batfishes. PC is CO-PI, PI is Matthew P. Davis (LSU), July 9, 2010.
$21,000	*Carl and Laura Hubbs Fellowship*, 2005–2006 stipend
$13,758	University of Michigan Graduate Student Grants from 2002–2006; *Donald W. Tinkle Scholarship, UMMZ Fish Division Grant, Hinsdale Scholarship Award, Rackham Discretionary Fund, Block Grant/Peter Okkelberg Awards, EEB/Rackham Travel Grants*

Recent Contributed and Invited Talks   *Presenter **Poster

Chakrabarty, P.* 2010.	Homology through history. BIOL 7111: Systematic Biology. September 14 [GUEST LECTURER]
Chakrabarty, P.* 2010.	Fishes!, East Baton Rouge Parish Goodwood Library. Children's Services (Elementary School Children), July 15 (INVITED TALK).
Chakrabarty, P.*, Albert, J.S. 2010.	Freshwater Fishes of Central America and a new take on the Great American Biotic Interchange. Joint Meeting of Ichthyologists and Herpetologists, Providence, Rhode Island. 9 July.
Chakrabarty, P.* 2010.	Biogeography of freshwater fishes: from the Greater Antilles to Gondwana. Biology Seminar Series, Southeastern Louisiana University, Hammond, April 22. (INVITED TALK)

Chakrabarty, P.* 2009.	Freshwater fishes and biogeography. Biology Seminar Series, Tulane University, New Orleans, September 25. (INVITED TALK)
Chakrabarty, P.* 2009.	Blind cave fishes from opposite ends of the Indian Ocean reveal an ancient Gondwanan connection. 8th Indo Pacific Fish Conference, Freemantle, Australia, June 2. (WINNER – Best Talk of Session)
Chakrabarty, P.*, Smith, W.L., Sparks, J.S. 2008.	Testing the Role of Sexual Selection in the Evolution of Ponyfishes. Weekly Seminar, Institute of Marine Biology, National Sun Yat-sen University, Kaohsiung, Taiwan, 16 November, 2008. (INVITED TALK)
Chakrabarty, P.*, Sparks, J.S. 2008.	The influence of sexual selection in the diversification of Leiognathidae. Joint Meeting of Ichthyologists and Herpetologists, Montreal, Quebec, Canada. 28 July 2008.

Recent Field Work

Winter 2010	*January 26–January 29, **Gulf of Mexico**. Trawled with SEAMAP (Southeast Area Monitoring and Assessment Program) on the R/V Pelican organized by the Louisiana Department of Wildlife and Fisheries. Collected 55 species for the LSU collection including, searobins, marine catfish, frogfish, batfish, puffers and flatfish. We also collected CTD-DO data, water samples, and sediment samples at 12 trawl stations.*
Fall 2010	*November 11–November 19, **Taiwan**. Made market collections of 260 marine species (including specimens from coral reefs and the deep sea) totaling over 1000 specimens. Traveled to markets in Taipei, Wuchi, Chiayi, Dongshih, Heng-chun, Tongkang and Tashi. Sampling included molecular and morphological specimens (360 lots) to be deposited at the Ichthyology Collection at LSU.*
Spring 2009	*March 30–April 9, **Sri Lanka**. Collected from over 20 sites on the western and southern Coasts of Sri Lanka, with lots of support from Thasun Amarasinghe and Rohan Pethiyagoda. The northern-most point visited was Puttalam (08°01'49.5"N, 079°49'42.0"E), the southern-most point Matara (05°56'50.2"N, 080°32'55.3"E) and the eastern-most was Habantota (06°07'27.2"N, 081°07'31.1"E). More than a dozen rare ponyfish species were collected as were many other marine and brackish water fishes.*

Curatorial Experience

2008–   *Collections.* Added more than 1000 lots and 5000 specimens to the LSU collection from fieldwork and gifts to the collection. The DNA tissue collection now contains more than 1000 samples representing nearly as many species.

2006–2008   *Lab Manager, Ichthyology Molecular Lab, American Museum of Natural History.* In charge of ordering materials (*e.g.*, tips, Qiagen kits, PCR beads, etc.), training students, and general upkeep of the Department of Ichthyology's molecular lab facilities at the American Museum of Natural History. I oversaw the move and initial set-up of this facility in the AMNH Molecular Systematics Lab complex. I was also trained to do (and occasionally called upon to do) daily and weekly maintenance and general upkeep of the ABI 3730XL sequencer.

2004, 2005   *Curatorial Assistant, Fish Division, University of Michigan Museum of Zoology.* My duties included: sorting and identifying fishes, x-raying, clearing and staining specimens, as well as day-to-day maintenance of museum specimens. I also initiated a system of cataloging and storing samples for DNA analysis. I was supervised by the Fish Division Collections Manager Doug Nelson and Museum Director Dr. William L. Fink.

Student Committees/ Mentoring

2007, 2008   *High School Science Research Program (HSSRP) Mentor.* Trained two students working in the Department of Ichthyology 4 hours a week, AMNH. Students help describe a new species, learned to use geometric morphometrics and molecular systematic techniques. Students prepared poster for JMIH meetings.

Teaching Experience

Fall 2010   Co-*instructor,* Curatorial Methods BIOL 7800 (Biological Sciences). Co-taught with other LSUMNS curators. Louisiana State University

Winter 2006   *Graduate Student Teaching Assistant,* for "Evolution and the Nature of Science", 6 two hour undergraduate workshops for the University of Michigan Evolution Theme Semester, January – March

Fall 2003–2005	*Graduate Student Instructor*, Graduate Student Instructor Training Course, gave lectures and trained incoming first time graduate instructors in Biology. University of Michigan
Fall 2003, 2004	*Graduate Student Instructor*, Chordate Anatomy and Phylogeny, University of Michigan
Winter 2002	*Graduate Student Instructor*, Special Topics in Ecology and Evolution, University of Michigan
Fall 2001, 2002	*Graduate Student Instructor*, Animal Diversity, University of Michigan

Committees

2008–2009	*President*, Graduate Researchers in Ecology and Evolutionary Biology (GREEBs). Created EEB Graduate Student, GREEBs' Website, organized meetings, typed up meeting minutes, created agenda and carried out requests. University of Michigan
2005–2003	*Graduate Student Representative*, Admissions Committee, UMMZ Curator's Meetings, and Diversity Committee. University of Michigan
Winter 2000	*President,* MacDonald Campus Student Society (MCSS), president of undergraduate and graduate student body, McGill University

Recent Service/Synergistic Activities

2010	*Presenter*, to elementary school children at the East Baton Rouge Parish Library. Spoke about Fish Diversity and the Gulf of Mexico Oil Spill. July 15
2010	*Judge*, Best student paper award in Ichthyology (Stoye Award) Joint Meeting of Ichthyologists and Herpetologists. Providence, Rhode Island. July 8–13.
2010	Panelist, Gulf Oil Spill Information Session at the Joint Meeting of Ichthyologist and Herpetologists, July 11 (Providence, Rhode Island)
2010	NSF workshop participant and white paper co-author for: New research opportunities emerging from integrating data across different taxonomic collections, March 25–27.
2010–	Encyclopedia of Life, Curator (www.eol.org)
2010–	Book Editor, *Copeia*
2010–	Assistant Editor, *Journal of Fish Biology*

2009–   *External manuscript reviewer,* for *Systematic Biology,
              Cladistics, Zoological Journal of the Linnean Society, Journal of
              Fish Biology, Copeia, Journal of Biogeography, Zootaxa,* BMC
              *Evolutionary Biology, Open Marine Biology Journal,
              Chromosome Research,* Nature Education Knowledge Project,
              PLoS One, Aquaculture, Grzimek's Animal Life
              Encyclopedia

Awards and Honors

2009–2013   Elected Member of the Board of Governors, American
              Society of Ichthyologists and Herpetologists (5 Yr. Term)
2006   *Carl and Laura Hubbs Fellow/Donald W. Tinkle Scholarship,
              University of Michigan Museum of Zoology,* for publication
              record and departmental service.
2005   *University of Michigan Outstanding Graduate Student
              Instructor Award,* (Highest honor given to GSIs under the
              auspices of the University of Michigan)
2000   *Valedictorian Class of 2000,* of the Macdonald Campus of
              McGill University

Appendix 7: Example Teaching Statement and Philosophy for Job Application

TEACHING STATEMENT AND PHILOSOPHY

Prosanta Chakrabarty

Teaching at the American Institute of Zoology would be a wonderful opportunity for any scientist. I would expect to be called upon to teach ichthyology and would very much look forward to doing this. My ichthyology class would cover topics that include an overview of fish diversity, physiology and comparative anatomy as well as research topics ranging from biogeography to phylogenetic analysis. I would also be willing to teach a graduate seminar on geometric morphometrics. Geometric morphometrics is a tool that allows us to quantitatively compare, contrast, and measure the magnitude of morphological diversity (GM techniques include Principal Components Analysis, Canonical Variate Analysis and analyses of disparity). If there is a need, I would be open to teaching any number of potential courses either at the graduate or undergraduate level.

I am proud of my teaching record at the University of Michigan, where I was awarded the Outstanding Graduate Student Instructor Award. (The most significant award a graduate student instructor can receive at that university.) I've taught courses in Animal Diversity, Parasitology, Chordate Anatomy, and a training course for first-time graduate student instructors (see CV for a complete list). I am enthusiastic about teaching and hope to inspire students to work in biology and natural history. I would take full advantage of my connections to the AIZ Museum of Natural History in teaching classes in the biology department. I would utilize the museum as a tool to teach a side of biology that many undergraduates might be unfamiliar with – particularly collections based research. I would very much like to mix lectures, fieldwork, and labwork in any class I teach.

My current postdoctoral position doesn't require teaching but I take advantage of opportunities to teach when I can. I have taught several

A Guide to Academia: Getting into and Surviving Grad School, Postdocs, and a Research Job, First Edition. Prosanta Chakrabarty.

high school students, undergraduates and graduate students working on projects with me on geometric morphometrics, alpha taxonomy, and molecular techniques (see Teaching Experience in CV). I have also talked to groups of undergraduates about career paths in the sciences and to donors about the importance of funding systematic research.

At the American Institute of Zoology I hope to have an active lab with graduate and undergraduate students working with me on various projects involving the ichthyology collection. I will train students in ichthyology, phylogenetic systematics, bioinformatics, computational biology and various other related fields. It is very important to me to train students with broad interests but also to have students working on projects that will utilize the ichthyology collections and emphasize the museum in their research.

Teaching Philosophy

I think a professor must be a pluralist, flexible in teaching styles in order to accommodate the class and reach every student. A teacher must get to know the students on an individual level and must learn each student's strengths and weaknesses. I try to take advantage of the variation in strengths among the students by varying the activities of the class. I like to set-up my classes so that my students feel like I am joining them as we try to reach the same goal. This is why I always incorporate active learning techniques that allow students to teach and learn from each other.

In class, I try to familiarize the students with the activities and goals during the lecture and have them periodically break into groups while I observe and give guidance. Group work allows the students to be teachers. If I see a number of groups struggling I change the teaching strategy. Instead of just lecturing and explaining, I demonstrate by putting myself in the role of a student 'that gets it.' For instance, if the activity is a difficult dissection, I will have the students observe me carrying out part of the dissection. I concentrate on the areas where the students are having the most trouble. Once I explain how to overcome the problem, I normally ask one of the students to continue with the dissection I started. This allows the emphasis to be on one of their peers. Observing a peer being successful often gives students confidence that they can follow suit.

Although most teachers set and discuss goals at the beginning of class, few follow-up on these goals at the conclusion of class. Before I let students leave (purging their memory as they go) I always review what the goals were and if they were met. I try to set each activity in the context of the entire class and larger subject area. This helps students see the point and often makes the activities more fulfilling and interesting.

Tying the students' motivations (*e.g.*, doing well in the course, having fun) to the teachers own (*e.g.*, expressing concepts clearly, reaching the goals of the experiments) makes for an exciting setting. Being excited about a subject is contagious. Students learn subjects better if they are interested in them.

I love the challenges and the rewards of teaching. My teaching philosophy of reaching every student with active learning techniques and flexible strategies constantly pays off and I often learn as many new things about the subject as my students.

Appendix 8: Example Extra Statement for Job Application (Curatorial Statement)

STATEMENT OF CURATORIAL INTERESTS AND EXPERIENCE

Prosanta Chakrabarty

As a museum trained systematist that works with specimens on a daily basis I care deeply about collections and curation. I think the model for a perfect collection is one that is continually expanding with the products of fieldwork in new areas while maintaining and taking advantage of morphological and molecular specimens (*e.g.*, cleared and stained, alcohol preserved, as well as tissue samples, extractions, *etc.*) so they may be studied forever. I think collections should be accessible to anyone around the world through loans and media accessible via the internet. Unsorted collections should be dealt with as quickly as possible with minimal backlog; staff ID days and the opinions of visiting experts should be used to deal with unidentified materials. I have found that sifting through unsorted material is not only necessary for the upkeep of the collections but also helps tune up my identification skills and in the discovery of new research problems.

As curator of ichthyology I will promote and build the collections. I will apply for grants to be used to modernize and expand the American Institutive of Zoology ichthyology collections, particularly the NSF funded Biological Research Collections Program. I have also discussed with Dr. Frank Hart (ichthyologist and Director of the Boulane University Museum) how the NSF funded Research Coordination Grant might improve communication between the ichthyological collections in the state, in country and abroad. I think this is an exciting time for museum collections given the number of major funds available for collections based research (*e.g.* AToL; Encyclopedia of Life, REVSYS); the opportunity should not be missed to enhance and modernize collections.

At the University of Michigan's Museum of Zoology, I had the opportunity to be the Fish Division's Curatorial Assistant for two terms. Besides learning about basic and important curation practices like cataloging and proper storage of specimens, I also left behind a lasting contribution. I started the curation of the UMMZ's Fish Division DNA collection. At the beginning of my graduate career there was no curation of tissue samples or extractions. I found this unfortunate because of the important role of molecular work in systematics. After a field season in the Dominican Republic I brought back tissues and vouchers from various fish groups. At this time the Fish Division DNA collection consisted of a -80 freezer with tubes labeled in the exclusive style of the individual researcher. There was no consistency, and none of the samples belonged to the UMMZ but rather to the individual researcher. These personal cataloging systems were often cryptic, so after that person stopped working on that project very little could be done to determine the origins of the sample. As the curatorial assistant I came up with a protocol for cataloging DNA and incorporating tissues and DNA extract into the UMMZ catalog. Besides organizing previous samples, I implemented a standard procedure of giving each sample its own tube with an individual number and position in a storage box that was linked to a voucher specimen in the UMMZ fish collections. This voucher was stored and clearly marked as a DNA voucher, comparable to how type specimens are labeled. This information is also stored in the on-line and in-house cataloging system at the UMMZ. Researchers can still catalog tissues for their own use but a sub-sample must also be deposited with the Fish Division. Since implementation of this program, researchers returning from the field not only deposit their specimens but also tissues from those specimens. This adds an entirely new dimension to the wonderful collections there.

Collections in general are our greatest record of the natural world. Besides the obvious need for collections in biological work, collections are great for grabbing the interest of the public, including donors. I often volunteered my time to give tours of the collections both at the UMMZ and at the AMNH. Showing someone a 200-pound coelacanth (*Latimeria chalumnae*) or an equally bizarre gulper eel (*Eurypharynx pelecanoides*) always left a lasting impression.

I have learned a great deal about how important public exhibits are to supporting research collections. Large research collections with poorly developed exhibits are often poorly funded and understaffed. Collections like the AMNH, FMNH, and Smithsonian with large exhibits are far better off because public awareness is raised in support of the biological collections.

Collecting is an essential part of my research program. I have collected in the Dominican Republic, Mexico, Belize, Singapore, Taiwan, Thailand, Malaysia, Indonesia and Sri Lanka bringing back hundreds of new specimens and species to museums there and in the U.S. Collecting always sparks new life into my research and making collections from marine and freshwater habitats throughout the world is not only a perk of our profession but a necessity. I understand that being a researcher at your institution comes with the responsibility of not only publishing top rate research and teaching but also bringing in and maintaining collections. I hope to follow in the footsteps of the many University of Michigan and AMNH graduates that have led major ichthyology collections. Important to me, as it was to them, is maintenance of a growing collection that is easily accessible to visiting and in-house researchers. I would strive to continue the great tradition of making important field collections, and advancing science and curation practices at the American Institute of Zoology Museum of Natural History.

Appendix 9: Example Chalk Talk for Job Interview

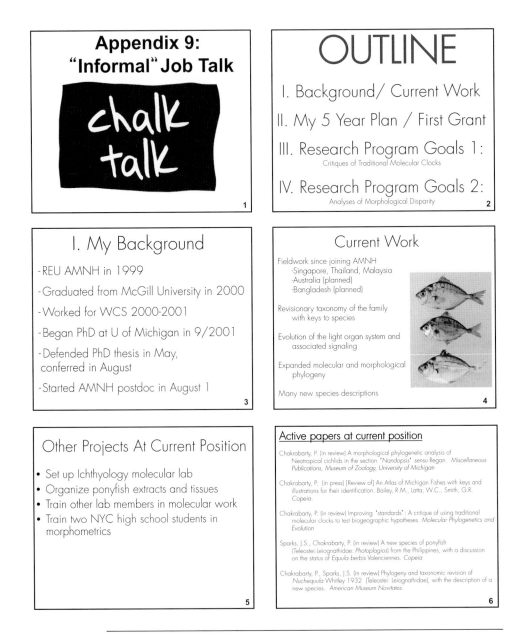

Appendix 9: "Informal" Job Talk

chalk talk

1

OUTLINE

I. Background/ Current Work

II. My 5 Year Plan / First Grant

III. Research Program Goals 1:
Critiques of Traditional Molecular Clocks

IV. Research Program Goals 2:
Analyses of Morphological Disparity

2

I. My Background

- REU AMNH in 1999

- Graduated from McGill University in 2000

- Worked for WCS 2000-2001

- Began PhD at U of Michigan in 9/2001

- Defended PhD thesis in May, conferred in August

- Started AMNH postdoc in August 1

3

Current Work

Fieldwork since joining AMNH
-Singapore, Thailand, Malaysia
-Australia (planned)
-Bangladesh (planned)

Revisionary taxonomy of the family
with keys to species

Evolution of the light organ system and
associated signaling

Expanded molecular and morphological
phylogeny

Many new species descriptions

4

Other Projects At Current Position

- Set up Ichthyology molecular lab
- Organize ponyfish extracts and tissues
- Train other lab members in molecular work
- Train two NYC high school students in morphometrics

5

Active papers at current position

Chakrabarty, P. (in review) A morphological phylogenetic analysis of
Neotropical cichlids in the section "Nandopsis" sensu Regan. Miscellaneous
Publications, Museum of Zoology, University of Michigan

Chakrabarty, P. (in press) [Review of] An Atlas of Michigan Fishes with keys and
illustrations for their identification. Bailey, R.M., Latta, W.C., Smith, G.R.
Copeia.

Chakrabarty, P. (in review) Improving "standards": A critique of using traditional
molecular clocks to test biogeographic hypotheses. Molecular Phylogenetics and
Evolution

Sparks, J.S., Chakrabarty, P. (in review) A new species of ponyfish
(Teleostei:Leiognathidae: Photoplagios) from the Philippines, with a discussion
on the status of Equula berbis Valenciennes. Copeia

Chakrabarty, P., Sparks, J.S. (in review) Phylogeny and taxonomic revision of
Nuchequula Whitley 1932 (Teleostei: Leiognathidae), with the description of a
new species. American Museum Novitates.

6

A Guide to Academia: Getting into and Surviving Grad School, Postdocs, and a Research Job,
First Edition. Prosanta Chakrabarty.
© 2012 John Wiley & Sons, Inc. Published 2012 by John Wiley & Sons, Inc.

II. 5 year Plan

- Get multiple NSF grants
  - REVSYS, PEET (for research component)
  - BRC (for collections component)
  - PEET (for teaching component)
- Publish 15 to 25 papers
  - Evolution of marine fishes
  - Taxonomy
  - Biogeography of Neotropical freshwater fishes

7

My First NSF Grants

Biogeography of Greater Antillean Freshwater Fishes

Limia, Gambusia, Poecilia, Lucifuga, Rivulus, etc.

Broader Impacts

Freshwater conservations of Lake Miragoane, and Haiti
The Madagascar of the West

Collaborators

Carlos Rodríguez, other Dominican colleagues

PEET - Partnerships for Enhancing Expertise in Taxonomy

REVSYS - Revisionary Syntheses in Systematics

8

My First NSF Grants – collections

BRC - Biological Research Collections (July, annually)

increase and improve space,

organize symposia,

frozen tissue collections,

digital images of types,

inventories accessible through the web, GIS

-French Museum

-Dominican Republic, Haiti

9

First Set of Future Papers

(1) Taxonomic revision of Cuban Cichlidae

(2) On a highly endangered radiation of *Limia* in Lake Miragoane, Haiti, with a phylogeny of Caribbean Poeciliids

(3) Historical biogeography of Cayman Ridge freshwater fishes

(4) Historical biogeography of Greater Antillean freshwater fishes

(5) On the absence of freshwater fishes from Puerto Rico

(6) On the absence of Ostariophysans from the Greater Antilles

(7) Relationships of the Galapagos hot-spot and the Greater Antilles

(8) Phylogeny and biogeography revision of *Lucifuga*

10

My other interests at my new position

1) Building and using the Neotropical collections

2) Building peripheral collections

3) Training students, postdocs, researchers from other countries

4) Outreach through teaching

11

III. Research Program 1

- Discuss your main anticipated research goals

IV. Research Program 2
Etc., etc.

- Discuss your other research goals

12

Appendix 10: Example Start-Up Wish List

Prosanta Chakrabarty/ Ichthyology Professor/Curator Position

Preliminary List of Equipment Required for Research Program
Budgetary Quotation

Catalog Number	Description	Quantity	Unit Price	Total Price	Comments
Molecular Laboratory Supplies					
Fisher 11686820	LIQUID NITROGEN CONTAINER- 20L PORTABLE	1	$1,262	$1,262	
Fisher 12-814-5Q	Thermo Scientific* MaxiMix* II Vortex Mixer	1	$410	$410	
Fisher 11-954-751	Guard-Rail Utility Cart	1	$604	$604	
Fisher 22-363-152	Nutating Mixer	1	$485	$485	
Fisher 0540285	PIPET 2100 SERIES 0.1 TO 2.5UL	2	$281	$561	
Fisher 0540287	PIPET 2100 SERIES 2 TO 20 UL	2	$281	$561	
Fisher 0540288	PIPET 2100 SERIES 10 T0 100UL	2	$281	$561	
Fisher 0540290	PIPET 2100 SERIES 100-1000UL	2	$281	$561	
Fisher 0540291	PIPET 2100 SERIES 500-5000UL	2	$281	$561	
Fisher 13688505	PIPET RESEARCH 12CH 30-300UL	1	$869	$869	
Fisher 213809	PIPET EPPENDORF REPEATER PLUS	1	$461	$461	
Fisher 11718	PIPET- BASE UNIT 8X9X31/2" 115V	1	$208	$208	
Fisher E5331000 045	PCR- Eppendorf* Mastercycler* Thermal Cyclers	1	$7,215	$7,215	
Fisher 04-977-146	Multichannel pipettes 0.5-10 size	2	$620	$1,240	
Fisher 03-840-7	Multichannel pipettes2.5-25UL size 2X ($745) - $1,490	2	$760	$1,520	
Fisher 03-840-29	Multichannel pipettes 20-200UL size 2X ($745) - $1,490	2	$760	$1,520	
Fisher FB-PCR2	Fisher Scientific PCR Workstation	1	$4,240.00	$4,240	
Fisher 97-935-1	Undercounter Refrigerator/Freezers	1	$2,398.00	$2,398	
USA Scientific 1110-3700	TipOne 0.1-10 ul	40	$18.00	$720	
USA Scientific 1111-4800	TipOne 0.5-10 ul	40	$29.60	$1,184	
USA Scientific 1110-8700	TipOne 1-300 ul	40	$19.50	$780	
ISC Bioexpress P-3288-1	1000UL Blue Pipet Tip Bulk 1000/pk	20	$13.00	$260	
Qiagen #69506	DNEasy Tissue Kit (250)	2	$504.00	$1,008	
Applied BioSystems	BigDye	20	$800.00	$16,000	
Agencourt 000132	Ampure 450ml Kit 12,500 (20 microliter PCR reaction)	4	$1,716.00	$6,864	
Invitrogen	Primers	12	$12.00	$144	
A. Daigger EF3274C Axygen®	Thermal Cycler Plates ,non-Skirted 96-Well, Clear, Case of 50,	3	$169.00	$507	
Daigger® EF4232A	Locking Microcentrifuge Tubes 1.5 Case of 5000,	3	$266.00	$798	
Genesee Scientific 22-555	96-well optical plates (axygen #PCR-96M2-HS-C) abi 3730	20	$36.00	$720	
Agencourt 000136	Agencourt CleanSEQ Kit (50 ml)	5	$1,716.00	$8,580	
Applied BioSystems (N8010560)	MicroAmp™ Optical 96-Well Reaction Plate	20	$44.00	$880	
Cambrex -51200	Molecular Biology Grade Water - 1 L	5	$24.00	$120	
Fisher Scientific 13-380-118	SHARPIE, BLK PERM XFINE 12/PK	2	$14.60	$29	
Fisher Scientific BP2573100	100BP DNA LADDER	2	$105.94	$212	
Fisher Scientific E0030127471	FOIL PCR ADHESIVE 100/PK	2	$130.50	$261	
Fisher Scientific BP1700-500	Proteinase K	1	$455.48	$455	
Fisher Scientific NC9803801	BIO-HAZARD BUCKET	1	$13.92	$14	
Fisher Scientific 14-432-22	Falcon TUBE SCREW CAP GRAD 500/CS	5	$104.06	$520	
Fisher Scientific (21-380-8J)	COMBI TIPS 0.05ML 50/PK	5	$118.68	$593	
Fisher Scientific (08-757-13A)	PETRI DISH 60X15 500/CS	5	$152.14	$761	
GE Healthcare	puRe Taq RTG PCR Beads (27955702)	15	$626.90	$9,404	
Invitrogen 11304-011	Platinum® Taq DNA Polymerase High Fidelity	2	$129	$258	
USA scientific 2310-5848	100-place hinged boxes, mixed neon colors	10	$28.50	$285	
Pharmco 111000200CSGL	Ethyl alcohol 200 proof, Absolute HPLC-UV	5	$102.00	$510	
ISC Bioexpress R-8100-1	100 Cryogenic Storage Rack	40	$4.85	$194	
			Subtotal	**$77,300**	

A Guide to Academia: Getting into and Surviving Grad School, Postdocs, and a Research Job,
First Edition. Prosanta Chakrabarty.
© 2012 John Wiley & Sons, Inc. Published 2012 by John Wiley & Sons, Inc.

Molecular Laboratory Supplies (Shared)

Catalog Number	Description	Quantity	Unit Price	Total Price	Comments
Fisher 15467547	FREEZER -86 C- 17CF DD ALRM -86C 115V	1	$8,576.50	$8,576.50	*
Fisher 15467441	FREEZER- STORAGE RACK- BOX 17/23FT	4	$146.20	$584.80	*
Fisher 15467449	FREEZER- STORAGE RACK- BOX 17/23 DD	4	$144.50	$578.00	*
	Lab supplies (glassware, tubing, tools, etc.)	1	$4,000.00	$4,000.00	Approx.
	Consumables (plasticware, culture, etc)	1	$4,000.00	$4,000.00	Approx.
	Reagents (chemicals, enzymes, isotopes)	1	$12,000.00	$12,000.00	Approx.
Applied BioSystems (4363935)	Pop-7	1	$3,630.00	$3,630.00	
DJBlabcare10732	Tabletop centrifuge Sigma 4-15C (Qiagen)	1	$7,407.34	$7,407.34	
Fisher E5331000 045	PCR- Eppendorf* Mastercycler* Thermal Cyclers	1	$7,215	$7,215	
	Lab Renovations		$50,000.00	$50,000.00	
			Subtotal	**$97,992**	

Office

Catalog Number	Description	Quantity	Unit Price	Total Price	Comments
Mac, DELL	Computers	2	$2,000.00	$4,000.00	
Overstock.com	Desk, chairs	3	$226.00	$678.00	
OfficeDepot 801811	HON® 1000 4-Drawer Letter File, Putty	3	$250.00	$750.00	
OfficeDepot Item #: 169560	Shelves	4	$659.00	$2,636.00	
Fisher 11-954-751	Guard-Rail Utility Cart	1	$603.84	$603.84	
www.genecodes.com	Sequencher Key/Sequencher	2	$2,500.00	$5,000.00	
More Direct K42751	Microsoft Office Mac/PC	1	$297.80	$297.80	
paup.csit.fsu.edu	PAUP*	1	$60.00	$60.00	
Everything Outlet	Adobe CS3	1	$635.00	$635.00	
Bluedoginc.com	Printer HP laserjet 1320	2	$388.99	$777.98	
www.cladistics.com/	TNT Phylogenetics	1	$45.00	$45.00	
			Subtotal	**$15,484**	

Field Work

Catalog Number	Description	Quantity	Unit Price	Total Price	Comments
	Taiwan 2 weeks, 2 people			$8,500	
	Mexico 2 weeks, 2 people			$4,000	
			Subtotal	**$12,500**	

Morphology Tools, Morphometrics

Catalog Number	Description	Quantity	Unit Price	Total Price	Comments
B&H Photo DSCH5B	SonyDSC-H7B Digital Camera 7.2 Megapixel	1	$384.95	$384.95	
B&H Photo SDMSPD-2048-A10	Sandisk Memory Stick 2GB	2	$44.75	$89.50	
B&H Photo	Nikon Zoom Lens f/4.0-f/32	1	$399.00	$399.00	
B&H Photo	Nikon Camera 10.2-Megapixel Digital SLR Camera Body - Black	2	$44.75	$89.50	
Ship the Web	Insect Pins #3 Black Insect Pins	5	$5.95	$29.75	
Dell.com	Dell Computer for Morphometrics	1	$2,000.00	$2,000.00	Approx.
Roboz	Microdissecting equipment			$2,500.00	Approx.
Fisher Scientific 12-640	BLADES SINGLE EDGE 100/PK	5	$23.43	$117.15	
Fisher Scientific 13-380-118	SHARPIE, BLK PERM XFINE 12/PK	2	$14.60	$29.20	
Amazon FWFL 160	Smartdisk 160gb	1	$144.94	$144.94	
SanDisk SDDR-108	Sandisk Memory Stick Reader	1	$19.90	$19.90	
B&H Photo MFG ULTRALUX	VELBON SDV-30 Tripod	1	$19.90	$19.90	
Morrell	Nikon SMZ Sterozoom Microscope w/camera, stand	1		$3,375.00	
Morrell	Nikon Microscope w/camera, stand	1		$14,292	
	X-Ray Machine	1	$27,500	$27,500	Approx.
	X-Ray Plates, Developer, Stopper			$300	Approx.
	Fish Books for Identifications			$1,000	Approx.
			Subtotal	**$52,291**	

Ichthyology Collections

Catalog Number	Description	Quantity	Unit Price	Total Price	Comments
Dell	Computer for Databasing collections	1	$2,000.00	$2,000.00	Approx.
Fisher 11-954-751	Guard-Rail Utility Cart	3	$603.84	$1,811.52	
CityChemicals	ETOH Drums	4	$450.00	$1,800.00	
	Temp/Humidity Ctrl			$10,000.00	Approx.
KOLS	Paragon Jars and Lids	10,000	$2.50	$25,000.00	Approx.
	Boxes Wooden	500	$20.00	$10,000.00	Approx.
CityChemicals	Formaldehyde			$800.00	Approx.
CityChemicals	Glycerine	1	$600.00	$600.00	Approx.
	Aluminum Tanks	20	$200.00	$4,000.00	Approx.
Staples	Cabinet for Cleared and Stained Material	2	$179.99	$359.98	
Memphis Net and Twine	Seine, 3 types	3	$75.00	$225.00	
	KOH			$200.00	Approx.

usplastic.com	Nalgene Jugs	assorted		$500.00	
Staples	Dymo Labelers	3	$35.00	$105.00	
Alpha Systems Inc	Label Printer Datamax W Class Printer	1	$2,850.00	$2,850.00	
Alpha Systems Inc	Network Interface Card 10/100 ethernet	1	$425.00	$425.00	
Alpha Systems Inc	Standard Cuttier Kit, Printerhead Cloths, Static Brushes	1	$780.00	$780.00	
Alpha Systems Inc	Datamax PGR Ribbons 4" X 1500'	12	$52.00	$624.00	
Alpha Systems Inc	Datamax Preservation Tags 4" X 500'	5	$290.00	$1,450.00	
CityChemicals	Rotenone	1	$800.00	$800.00	Approx.
drillspot.com	Electronic Calipers	2	$469.39	$938.78	
ULINE S-2392	Polytubing 7" x 1,500'	2	$89.00	$178.00	
ULINE H-293	Heatsealer	1	$175.00	$175.00	
			Subtotal	**$65,622**	

| Miscellaneous /Personnel | | | | | |
Catalog Number	Description	Quantity	Unit Price	Total Price	Comments
	2 YR Research tech / collections manager	2 yrs.	$30,000.00	$60,000.00	+ benefits
	Postdoc	2 yrs.	$39,000.00	$78,000.00	+ benefits
	One semester teaching release		$0.00	$0.00	
	Summer-salary	2m/2 yr.	$28,000.00	$28,000.00	(at 7K a month)
	Student Discresionary Funds (travel, computer, supplies)	2 yr.	$4,000.00	$8,000.00	
	Moving Expenses			$5,000.00	
	Cobra Health Care (cover myself and wife)	2	$333.86	$667.72	
	DC 37 Prescription Drug Plan	2	$83.42	$166.84	
			Subtotal	**$179,835**	

Total Equipment List	**$501,023**

Without Shared	**$403,032**

Appendix 11: Review Package Summary Presentation

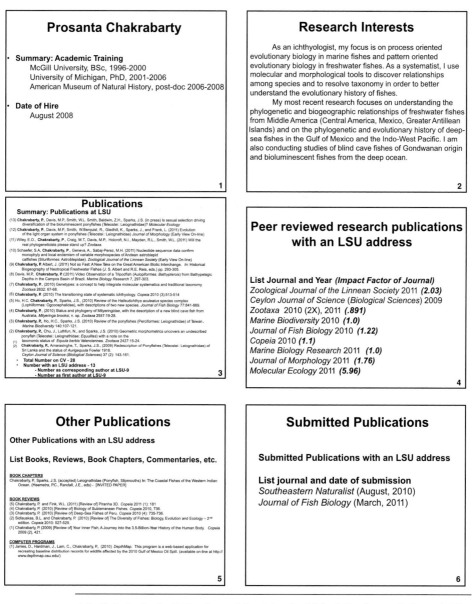

Prosanta Chakrabarty

- **Summary: Academic Training**
  McGill University, BSc, 1996-2000
  University of Michigan, PhD, 2001-2006
  American Museum of Natural History, post-doc 2006-2008

- **Date of Hire**
  August 2008

1

Research Interests

As an ichthyologist, my focus is on process oriented evolutionary biology in marine fishes and pattern oriented evolutionary biology in freshwater fishes. As a systematist, I use molecular and morphological tools to discover relationships among species and to resolve taxonomy in order to better understand the evolutionary history of fishes.

My most recent research focuses on understanding the phylogenetic and biogeographic relationships of freshwater fishes from Middle America (Central America, Mexico, Greater Antillean Islands) and on the phylogenetic and evolutionary history of deep-sea fishes in the Gulf of Mexico and the Indo-West Pacific. I am also conducting studies of blind cave fishes of Gondwanan origin and bioluminescent fishes from the deep ocean.

2

Publications

Summary: Publications at LSU

(13) Chakrabarty, P., Davis, M.P., Smith, W.L. Smith, Baldwin, Z.H., Sparks, J.S. (in press) Is sexual selection driving diversification of the bioluminescent ponyfishes (Teleostei: Leiognathidae)? *Molecular Ecology*
(12) Chakrabarty, P., Davis, M.P., Smith, W.Berquist, R., Gledhill, K., Sparks, J., and Frank, L. (2011) Evolution of the light organ system in ponyfishes (Teleostei: Leiognathidae) *Journal of Morphology* (Early View On-line)
(11) Wiley, E.O., **Chakrabarty, P.**, Craig, M.T, Davis, M.P., Holcroft, N.I., Mayden, R.L., Smith, W.L. (2011) Will the real phylogeneticists please stand up? *Zootaxa.*
(10) Schaefer, S.A, **Chakrabarty, P.**, Geneva, A., Sabaj-Perez, M.H. (2011) Nucleotide sequence data confirm monophyly and local endemism of variable morphospecies of Andean astroblepid catfishes (Siluriformes: Astroblepidae). *Zoological Journal of the Linnean Society* (Early View On-line)
(9) **Chakrabarty, P.** Albert, J. (2011) Not so Fast: A New Take on the Great American Biotic Interchange. *In:* Historical Biogeography of Neotropical Freshwater fishes (J. S. Albert and R.E. Reis, eds.) pp. 293-305.
(8) Davis, M.P., **Chakrabarty, P.** (2011) Video Observation of a Tripodfish (Aulopiformes: *Bathypterois*) from Bathypelagic Depths in the Campos Basin of Brazil. *Marine Biology Research* 7, 297-303.
(7) **Chakrabarty, P.** (2010) Genetypes: a concept to help integrate molecular systematics and traditional taxonomy. *Zootaxa* 2632: 67-68.
(6) **Chakrabarty, P.** (2010) The transitioning state of systematic ichthyology. *Copeia* 2010 (3):513-514
(5) Ho, H.C. **Chakrabarty, P.**, Sparks, J.S., (2010) Review of the *Halieutichthys aculeatus* species complex (Lophiiformes: Ogcocephalidae), with descriptions of two new species. *Journal of Fish Biology* 77:841-869.
(4) **Chakrabarty, P.**, (2010) Status and phylogeny of Milyeringidae, with the description of a new blind cave fish from Australia, *Milyeringa brooksi*, n. sp. *Zootaxa* 2557:19-28.
(3) **Chakrabarty, P.**, Ho, H.C., Sparks, J.S. (2010) Review of the ponyfishes (Perciformes: Leiognathidae) of Taiwan. *Marine Biodiversity* 140:107-121.
(2) **Chakrabarty, P.**, Chu, J., Luthfun, N., and Sparks, J.S. (2010) Geometric morphometrics uncovers an undescribed ponyfish (Teleostei: Leiognathidae: *Equulites*) with a note on the
  (1) taxonomic status of *Equula berbis* Valenciennes. *Zootaxa* 2427:15-24.
(2) **Chakrabarty, P.**, Amarasinghe, T., Sparks, J.S., (2009) Redescription of Ponyfishes (Teleostei: Leiognathidae) of Sri Lanka and the status of *Aurigequula* Fowler 1918.
  Ceylon Journal of Science (Biological Sciences) 37 (2): 143-161.
- **Total Number on CV - 28**
- **Number with an LSU address - 13**
  - **Number as corresponding author at LSU-9**
  - **Number as first author at LSU-9**

3

Peer reviewed research publications with an LSU address

List Journal and Year *(Impact Factor of Journal)*
Zoological Journal of the Linnean Society 2011 *(2.03)*
Ceylon Journal of Science (Biological Sciences) 2009
Zootaxa 2010 (2X), 2011 *(.891)*
Marine Biodiversity 2010 *(1.0)*
Journal of Fish Biology 2010 *(1.22)*
Copeia 2010 *(1.1)*
Marine Biology Research 2011 *(1.0)*
Journal of Morphology 2011 *(1.76)*
Molecular Ecology 2011 *(5.96)*

4

Other Publications

Other Publications with an LSU address

List Books, Reviews, Book Chapters, Commentaries, etc.

BOOK CHAPTERS
Chakrabarty, P, Sparks, J.S. (accepted) Leiognathidae (Ponyfish, Slipmouths) In: The Coastal Fishes of the Western Indian Ocean. (Heemstra, P.C., Randall, J.E., eds) - [INVITED PAPER]

BOOK REVIEWS
(5) Chakrabarty, P. and Fink, W.L. (2011) [Review of] Piranha 3D. *Copeia* 2011 (1): 181
(4) Chakrabarty, P. (2010) [Review of] Biology of Subterranean Fishes. *Copeia* 2010, 736.
(3) Chakrabarty, P. (2010) [Review of] Deep-Sea Fishes of Peru. *Copeia* 2010 (4): 735-736.
(2) Sidlauskas, B.L. and Chakrabarty, P. (2010) [Review of] The Diversity of Fishes: Biology, Evolution and Ecology – 2nd edition. *Copeia* 2010: 527-529.
(1) Chakrabarty, P. (2009) [Review of] Your Inner Fish, A Journey into the 3.5-Billion-Year History of the Human Body. *Copeia* 2009 (2), 421.

COMPUTER PROGRAMS
(1) Janies, D., Hardman, J., Lam, C., Chakrabarty, P., (2010) DepthMap. This program is a web-based application for recreating baseline distribution records for wildlife affected by the 2010 Gulf of Mexico Oil Spill. (available on-line at http://www.depthmap.osu.edu/)

5

Submitted Publications

Submitted Publications with an LSU address

List journal and date of submission
Southeastern Naturalist (August, 2010)
Journal of Fish Biology (March, 2011)

6

Research Funding

Research Funding

List
Grants obtained at LSU in which the candidate is the PI.

Reconstructing Heroini (Teleostei:Cichlidae) – Of Heroes,
Convicts, Angels and Red Devils,

National Science Foundation, $ 519,944

7

Research Funding

Research Funding

List

Grants obtained at LSU in which the candidate is a co - I.

Title	Agency	Name of PI	Total Awarded

8

Research Presentations

Research Presentations

Number of Invited Talks - 17

17 -LSU School of the Coast & Environment, April 1, 2011
16 -University of Southern Mississippi, Hattiesburg. October 22, 2010
15 -East Baton Rouge Parish Goodwood Library, July 15, 2010
14 -Joint Meeting of Ichthyologists and Herpetologists, Providence, Rhode Island. July 9, 2010
13 -Zachary High School, April 29, 2010
12 -Southeastern Louisiana University, Hammond, April 22, 2010
11 -NSF Collections RCN Data Integration Workshop, New Orleans, March 26, 2010
10 -Case Western University, Cleveland, Ohio March 18, 2010
9 -Sam Houston State University at Huntsville, November 21, 2009
8 -University of Louisiana at Lafayette, November 6, 2009
7 -Tulane University, New Orleans, September 25, 2009
6 -Joint Meeting of Ichthyologists and Herpetologists, Portland, Oregon. July 26, 2009
5 -Indo Pacific Fish Conference, Freemantle, Australia, June 2, 2009
4 -Renewable Natural Resources Seminar Series, LSU RNR, April 31, 2009
3 -Saturday Science Lecture Series, LSU Nicholson Hall, April 18, 2009
2 -Willi Hennig Society, Tucuman, Argentina, October 29, 2008
1-National Sun Yat-sen University, Kaohsiung, Taiwan, November 16, 2008

9

Poster Presentations

Poster Presentations

Matamoros, W., Davis, M.P., McMahan, C., Chakrabarty, P.,
2011. Estimating Divergence Times, Diversification Patterns
and Historical Biogeography of the New World Subfamily
Poeciliinae (Cyprinodontiformes: Poeciliidae). Joint Meeting
of Ichthyologists and Herpetologists, Minneapolis, Minnesota,
10 July.

10

Teaching

Summary

Course Title	Semester	# of Students	Credit Hours	Eval. (Q1-8) Eval. (Q9: Overall Inst.) Eval. (Q10: Overall Course)
Ichthyology BIOL/RNR 4145	Fall 2009	9	4.0	3.89 (College Avg 3.58) 3.88 (College Avg 3.43) 3.75 (College Avg 3.39)
Systematics Discussion Group BIOL 7091	Fall 2009	6	1.0	evaluations are not taken for seminar classes
Ichthyology BIOL/RNR 4145	Fall 2010	17	4.0	3.77 (College Avg 3.60) 3.85 (College Avg 3.44) 3.77 (College Avg 3.42)
Systematics Discussion Group BIOL 7091	Fall 2010	7	1.0	evaluations are not taken for seminar classes
Curatorial Methods BIOL 7800*	Fall 2010	7	1.0	*team taught, graduate level

11

Undergraduate Teaching

Undergraduate Teaching
Parker House (2011-), Lee Shelton (2011-), Justin Kutz (2009-)
Paige Patterson (2009) Maladah Bader (2009), Shelley Chauvin (2009), Jessica Salter (2009), Jill Dowling (2010)
Awards or other recognition

Grants funded for teaching
$ 117,531 Board Of Regents, Traditional Enhancement Grant - Education:
    Making a Big Splash with Louisiana Fishes: A Three-tiered
    Education Program and Museum Exhibit. PC is CO-PI, PI is Dr. Sophie Warny

Student accomplishments (publications, awards, etc.)
1 -Justin Kutz started as a Zachary HS student accepted into the LA STEM program at
    LSU, created www.cacichlids.com which hosts products from my NSF grant
2 - Lee Shelton will be a co-author on a paper in prep
3 -2 authors in bold are New York HS students
    Chakrabarty, P., Chu,J., Luthfun,N., and Sparks, J.S. (2010) Geometric morphometrics uncovers an undescribed ponyfish
    (Teleostei Leiognathidae:Equulites) with a note on the taxonomic status of Equula berbis Valenciennes.
    Zootaxa 2427:15-24.

12

Graduate Teaching

Graduate Teaching
Caleb McMahan (Ph.D.), Wilfredo Matamoros (Post-Doc), Matthew Davis
    (Post-Doc), Valerie Derouen (Masters of Natural Science, starting
    summer 2011)
Number of graduate students
    Caleb McMahan (Ph.D., started in Fall 2010)
Number of theses completed 0

Awards or other recognition
Student accomplishments (publications, awards, etc.)
    Caleb has 6 publications (4 with LSU addresses)

13

Departmental Service

List year and committee assignments

List the number of graduate student committees
    4 Ph.D.
    Caleb McMahan (2010-)
    Hector Urbina (2009-)
    Christopher Reed (2010-)
    Lorelei Patrick (2009-2011, no longer on committee)

14

College and University Service

List year and committee assignments
List the number of graduate student committees 1
(Scott Reed, Department of Comparative Biomedical Sciences)
Other

15

Service outside the university community

Service to the Discipline
2010-   Assistant Editor, *Journal of Fish Biology*
2010-   Book Editor, *Copeia*
2009-   Member of the Editorial Board of *Copeia*
2009-2014   Governor, American Society of Ichthyologists and
               Herpetologists
Service to the Local Community
-Presented *Saturday Science at LSU*, MNS Special Saturday Presenter, Ocean
    Commotion, public library talks, various other local, public talks....
-Mentor, (ZAAR) Zachary High School Student, Justin Kutz, 2009-2010
-Tiger HATS (*Human Animal Therapy* Service) volunteer at Parker House for
    abused and neglected children (own 3 therapy dogs)

16

Glossary

Academia The field and community of higher education and research at universities and other equivalent institutions

Administrators The staff of deans, directors, and chairs who oversee the faculty. The administrators may be faculty members themselves.

Assistant Professor A professor who does not have tenure. The first stage of a professorship. The assistant professor must be reviewed and pass tenure to become an associate professor.

Associate Professor A tenured professor. The second stage of a professorship. The associate professor must be reviewed before he can become a full professor.

Biographical Sketch A short summary of your CV that is often requested in grant proposals. This sketch often includes your educational background, selected publications, and a list of your collaborators, among other items.

Broader Impacts A statement that explains the broader social impact of your scientific work that is often requested in grant proposals. These social impacts can include anything from creating an exhibit at a museum to teaching elementary school children about science as it is related to your proposal.

Candidacy A Ph.D. student must pass some qualifying exams to become a Ph.D. candidate. A student is a "pre-candidate" until he passes these exams. Being a candidate allows a student to apply for additional funding sources, but it also means that the student is in good standing with the department and on his way to completing his degree.

Chalk Talk A somewhat informal presentation that some job candidates have to give. This talk often is a question–answer period during which the candidate can answer questions using a chalkboard (or whiteboard) as a tool. These can often include a short Powerpoint presentation as well.

Cohort A group of students who entered a degree program in the same year.

Competing Offer A job offer from a second institution that a job candidate can use as leverage to get an improved offer from another institution.

Corresponding Author The author of a scientific publication to whom correspondence (questions about the paper) is addressed.

A Guide to Academia: Getting into and Surviving Grad School, Postdocs, and a Research Job,
First Edition. Prosanta Chakrabarty.
© 2012 John Wiley & Sons, Inc. Published 2012 by John Wiley & Sons, Inc.

Cover Letter An introductory letter from a job applicant that is typically the first part of a job application.

CV short for *curriculum vitae*. It is a resume and list of experiences and training.

Defense A thesis defense is a presentation and a meeting with your graduate committee during which you "defend" (argue and justify) your thesis.

Dissertation see **Thesis**

Doctorate A Ph.D.

Exit Seminar The seminar given as part of a graduate defense.

Faculty The assistant, associate, and full professors in a department. The faculty may also include lecturers, in which case the professors that can have graduate students are know as the "graduate faculty."

Federally Funded Grant A grant that is funded through one of the national government granting agencies, such as the National Science Foundation or the National Institute of Health.

First Author The lead author on a publication, usually the person who did the bulk of the writing.

Full Professor A tenured professor in the final and highest level of a professorship.

General Exams An expression sometimes used instead of qualifying exams that are the exams that a student must pass to become a candidate.

GPA Grade Point Average. Your average grade in your classes, usually expressed as a number between 0 and 4.

GRE Graduate Record Examination. A standardized test that you usually must take before applying to graduate schools.

Grants Funding received from an agency after a submitted proposal is approved.

H-Index A measure of a researcher's publication quality and productivity. Calculated by counting the highest number of publications that have been cited a given number of times. If you have twenty publications and six were cited at least six times, your H-index is six, given that the remaining publications have less than six citations (see Figure 7.2).

Indirect Costs The grant money that an institution keeps for processing and handling of the grant, including administrative costs.

In Press A term used to describe a paper that is accepted for publication but not yet published.

In Prep. A term used to describe a paper that is being prepared. (Prep is short for preparation.) The paper has not yet been submitted.

In Review A term used to describe a paper that has been submitted for publication and is in the process of being reviewed by editors or outside reviewers.

Internal Grants Grants available from within your department or institution reviewed by people at the institution.

Job Line Allocated funds specifically for a certain academic position. For example, Harvard has a dedicated line to hire an ichthyologist. If their current ichthyologist leaves, they will be able to hire someone into that line.

Job Talk A research presentation given by a candidate for an open position.

Journal Club A class, either informal or formal, in which students and faculty discuss a paper from a journal.

Junior Faculty Assistant professors and other untenured faculty.

Junior Position An academic position advertised for a pre-tenured (Assistant) but tenure-track professorship.

Lab This term pertains to the physical research space of a PI or the PI's staff, including his graduate students, technicians, and postdocs.

Labmates The congenial term used to describe the other members of your lab aside from the PI; and usually just in reference to the other students other than yourself.

Lecturer A person, usually with a Ph.D., hired specifically to teach a class but usually without lab or research funds. Lecturers are typically not hired into a tenure-track position.

Line see **Job Line**

Long Short-List A list of potentially suitable applicants selected by a search committee for an academic position. This list is usually then narrowed down further by a vote from an entire department creating the "short-list" of applicants who will be interviewed for the position.

Master's Degree A graduate degree that usually takes 2–3 years to complete. This can be a thesis or non-thesis degree that is one step above a bachelor's degree but below the Ph.D. This degree can be a stepping stone for obtaining a Ph.D., but it is not required for entering a Ph.D. program.

Meeting see **Society Meetings**

NIH National Institute of Health. One of the largest granting agencies and, along with the NSF, one of the most prestigious.

NSF National Science Foundation. One of the largest granting agencies and, along with the NIH, one of the most prestigious.

Office Hours A dedicated time to meet with undergraduate student to discuss their class and to go over material from that class or lab.

Open Rank Position An academic job posting that is open to candidates of any rank (assistant, associate, or full professor).

Overhead see **Indirect Costs**

Peer Review The process during which your submitted manuscripts for publication are reviewed by peers in your field.

Perspectives Graduate student applicants who are being evaluated during a recruitment period.

Ph.D. Doctor of Philosophy. The highest degree obtainable for a given scientific subject. With this degree, you can become a professor at a university or a research scientist. This degree is always a thesis-based degree for which you must write a doctoral thesis based on your unique discoveries and results based on research you conducted for that degree.

PI Principal Investigator, the head of a lab, or main investigator on a grant.

Postdoc Postdoctoral Fellow or Associate, a person with a Ph.D. who has obtained a paid position or a fellowship.

Pre-candidate A Ph.D. student who has not yet passed candidacy (qualifying) exams.

Preliminary Exams (Prelims) see **Qualifying Exams**

Presentation A research or teaching talk given to a department or lab group in some academic setting.

Rebuttal Letter A cover letter that responds to criticisms or comments made by reviewers on a manuscript that is being returned as a revision to a journal.

Recommendation Letter A letter, typically required, for a job candidate or student written by their mentors or peers extolling that student's or candidate's abilities.

Recruitment Weekend A period of evaluation for perspective graduate students, during which they are being interviewed and eventually selected for admittance into a graduate program.

References The peers and mentors who would be willing to write recommendation letters on your behalf. (The term can also be used to mean citations for a scientific paper.)

Research Program A term used to describe your particular research interests and the scientific tools you are using to better understand that field of research.

Research Statement Part of a job application that describes your research program and future research interests.

Research I University A descriptive term to describe universities who conduct extensive amounts of research. At these institutions, most of the academics focus on research, usually with less priority on teaching and other duties.

Review Paper A scientific paper written with a focus on some aspect of some field that references and reviews most of the previous scientific works on that topic.

Search Committee A committee of faculty chosen to hire a new faculty member. This committee may write the advertisement for the position, select among the applicants, and make a recommendation for a hire, among other duties.

Seminar A talk or presentation, usually 45–50 minutes when given as part of a weekly series or 10–15 minutes if given at a society meeting with other talks in a symposium.

Senior Author The last author listed on a by-line of authors on a scientific paper usually reserved for the PI.

Senior Position An academic position advertised for tenured (Associate or Full) professors.

Service Extracurricular work done for your department, institution, or some external society, including volunteer work or being part of a committee.

Short List The final list of candidates (from among all applicants) that will be brought in for interviews for a job.

Society Meetings Meetings, usually annually, of a scientific society. Research presentations often make up the bulk of these meetings.

Soft Money Grant money or other nonguaranteed funds. A person working on soft money garners their salary and research funds from nonguaranteed funds.

Start-Up Package The funds given to new professors by their institutions to spend on getting their research programs going. This can include funds to buy lab equipment or to hire scientific staff (e.g. postdocs).

Summer Salary Additional salary not part of your guaranteed salary that can be requested in a grant.

Synergistic Activities Activities conducted as a volunteer or to advance the societal benefits of your profession and research (e.g., giving a talk at a public library to elementary school children about your research).

TA Teaching Assistant. Usually a graduate student who is teaching a lab portion of a class or assisting the professor in other aspects of the class.

Team Taught (Team Teaching) Classes taught by more than one professor. Some universities will count a team-taught class the same as a class where there is only one professor when reviewing someone's teaching efforts.

Teaching Abstention A break from teaching; you might ask for a teaching abstention when you start a new job to focus on other commitments.

Teaching Philosophy A statement often required in job applications to professorial positions explaining the candidates thoughts on teaching strategy.

Teaching Presentation A requirement some job searches have where a job candidate must give a presentation as if for one of the classes they may be asked to teach.

Tenure If you get tenure in a position, you cannot be fired without just cause. You are essentially granted your position for life or until you choose to retire.

Tenure Clock A descriptive term to describe the period of time between starting a tenure-track job and being reviewed for tenure.

Tenure Track The course you take from an assistant professorship to full professorship. A tenure-track position is a position where you must receive tenure to continue in that position.

Thesis The formal manuscript you write up explaining your graduate research.

Thesis proposal A written proposal outlining your ideas for your graduate thesis.

Two-Body Problem The phrase often used to describe having a spouse or significant other in an academic position when one or both partners are looking for a permanent position.

Qualifying Exams The exams you must take to achieve "candidacy" and be in good standing as a graduate student.

Index

DATE			

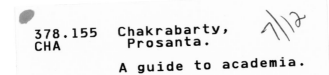